ENERGY: Global Prospects 1985-2000

Report of the Workshop on Alternative Energy Strategies

A Project Sponsored by the Massachusetts Institute of Technology

Carroll L. Wilson
Project Director

McGraw-Hill Book Company

NEW YORK ST. LOUIS SAN FRANCISCO

Auckland Bogotá Düsseldorf Johannesburg London Madrid Mexico Montreal
New Delhi Panama Paris São Paulo Singapore Sydney Tokyo Toronto

This material is based upon research supported in part by the National Science Foundation under Grant No. OEP74-13902 AO1. Any opinions, findings, conclusions or recommendations herein are those of the authors and do not necessarily reflect the views of the NSF.

Library of Congress Cataloging in Publication Data

Workshop on Alternative Energy Strategies.

Energy: Global Prospects, 1985-2000.

1. Power resources. 2. Energy policy. I. Wilson, Carroll L. II. Title.

TJ163.2.W653 1977 333.7 77-4700

ISBN 0-07-071878-4

ISBN 0-07-071879-2 pbk.

Designed & Printed at The Nimrod Press, Boston.

TABLE OF CONTENTS

Page

Page

Page

Page

Page

FOREWORD

The Workshop on Alternative Energy Strategies (WAES) was an experiment in international collaboration. Its unique form, membership, and style of operation have served well for our study of an important global subject—the problems of assuring a continuous flow of energy in the massive quantities all modern societies need for continuity—indeed for survival. The energy problem is a mixture of technical, economic and political factors, in each nation and globally. The critical interdependence among nations centers on oil, which now furnishes more than half of the national energy needs of industrialized societies.

WAES was born of my belief that more effective mechanisms were needed to study critical global problems. I selected energy as the study topic because I believed that the world may be moving steadily and with little apparent concern toward a new and massive energy crisis.

A time horizon to at least the year 2000 seemed necessary; freedom to focus attention on long-range issues—rather than on the short-range decisions that demand so much attention of governments and intergovernmental bodies—was essential. In addition, an ad hoc project lasting 2-3 years would not conflict with existing institutions. Because the Workshop was designed as a private and informal study group, government officials could take part—as individuals—together with others from industry, finance and academia.

The group had to be small and informal so that the Participants could engage in free-ranging discussion and work together to reach agreement on a plausible range of global energy futures and their implications for national policy and international cooperation. Such persons should come from countries that use most of the world's energy. Each person had to be free to represent his own views rather than those of his organization. Each should bring a wide range of personal experience and a network of national and international relationships that would allow him to communicate the Workshop's findings and their implications for energy policy in his own country and in international forums.

To achieve such a set of goals required a number of decisions for which I alone am responsible. I decided to select the 30 Participants principally from 15 industrialized countries that used 80% of the world's energy in 1972; three of these countries—Iran, Mexico, and Venezuela—are also important oil producers and exporters. Perceptions in the industrialized countries of the energy situation over the next 25 years and the actions they take to avoid or to diminish prospective shortages of fuel will largely determine the choices available to the rest of the world.

When I set out three years ago to find individuals who could participate fully, I decided to ask each Participant (senior member) to choose an Associate who would devote much of his time to the Workshop under the direction of the Participant. The intensity of the Associates' work has made this arrangement an indispensable feature of WAES. Associates met 13 times for 1-2 weeks over a period of 28 months. Participants met seven times for 2-4 days at a time. Each Participant agreed to arrange for the financing of all the work done at the national level and for the cost of his Associate's travel, time and other expenses. English was accepted as the Workshop language. Finally a small, capable, versatile, and dedicated staff at MIT was indispensable in leading and coordinating this global project within demanding deadlines. Generous grants from American sources made it possible to meet the costs of the MIT staff and their worldwide activities.

Has this experiment in international collaboration been successful? If successful, might a similar design be suitable for study of other global problems such as food and mineral resources? The Workshop has been effective in harnessing the efforts of a diverse, multinational group to develop a methodology for the study of global energy problems, to carry out analyses based on such methodology and to reach the conclusions reported in this volume and in the three technical volumes. We have learned that when responsible people from different countries work together as individuals on a common problem they can cooperate effectively in the kind of informal setting we have had in the Workshop.

Only time will answer questions about the scope and depth of our findings and their influence on perceptions of energy futures and actions to avert potential shortages. I believe our primary purpose has been to educate a wide and influential public. Without public understanding of the energy problems that lie ahead and the necessity for costly and sustained commitment to solve these problems, political leaders will lack support for their decisions and actions. The real measure of the Workshop's success lies in the clarity and persuasiveness of its findings and in its effectiveness in generating understanding and commitment among the people and their leaders.

My conclusion at the end of this Workshop is that world oil will run short sooner than most people realize. Unless appropriate remedies are applied soon, the demand for petroleum in the non-Communist world will probably overtake supplies around 1985 to 1995. That is the maximum time we have: thirteen years, give or take five. It might be less. Petroleum demand could exceed supply as early as 1983 if the OPEC countries maintain their present production ceilings because oil in the ground is more valuable to them than extra dollars they cannot use. We don't have much time to learn how to replace, or decrease our dependence on the fuel that for three decades has fed the expansion of Western living standards and the hopes of all nations for material betterment. Time is our most precious resource. It must be used as wisely as energy.

Carroll L. Wilson

WORKSHOP PARTICIPANTS, ASSOCIATES, AND STAFF

PROJECT DIRECTOR
Carroll L. Wilson
Massachusetts Institute of Technology

PARTICIPANTS

Canada
Mr. Marshall A. Crowe
Chairman, National Energy Board

Mr. Maurice F. Strong
Chairman, Petro-Canada

Denmark
Professor Bent Elbek
Niels Bohr Institute

Finland

Professor Jorma Routti
Helsinki University of Technology

France
M. Jean Couture
Conseiller du Président
Société Générale

M. Jean-Marie Martin
Director, Institut Economique et Juridique de l'Energie
Centre National de la Recherche Scientifique

Germany
Director Dr. Hans Detzer
Head of Central Planning
Badische Anilin & Soda Fabrik AG

Professor Heinrich Mandel
Member of the Board of Directors
Rheinisch-Westfälisches Elektrizitätswerk AG

Professor Hans K. Schneider
Director, Institute of Energy Economics
University of Cologne

Iran
Dr. Khodadad Farmanfarmaian

Italy
Professor Umberto Colombo
Director, Research & Development Division
Montedison

Professor Sergio Vaccà
Director, Istituto di Economia delle Fonti de Energia
Università L. Bocconi, Milano

Japan
Mr. Toshio Doko
President, Japan Federation of Economic Organizations (Keidanren)

Mr. Shuzo Inaba
President, Industrial Research Institute

Mr. Soichi Matsune
Chairman, Committee on Energy
Japan Federation of Economic Organizations (Keidanren)

Mr. Tatsuzo Mizukami
President, Japan Foreign Trade Council, Inc.

Dr. Saburo Okita
President, The Overseas Economic Cooperation Fund

Mr. Masao Sakisaka
President, National Institute for Research Advancement

Mr. Shigefumi Tamiya
Advisor to the Chairman
Enrichment and Reprocessing Group

Mexico
Ing. Juan Eibenschutz
Executive Secretary
Mexican National Energy Commission

Dr. Victor Urquidi
President, El Colégio de México

The Netherlands
Dr. A. A. T. van Rhijn
Deputy Director General for Energy
Ministry of Economic Affairs

Netherlands/U.K.
Mr. J. C. Davidson
Former Director, Shell International Petroleum Company, Ltd.
Royal Dutch/Shell Group

Norway
Mr. Christian Sommerfelt
Chairman
Elkem-Spigerverket A/S

Sweden
Mr. Erland Waldenström
Chairman
Gränges AB

United Kingdom
Mr. Robert Belgrave
Policy Adviser to the Board
British Petroleum Company, Ltd.

Professor Sir William Hawthorne
Master, Churchill College
University of Cambridge

United States
Mr. Thornton F. Bradshaw
President, Atlantic Richfield Company

Mr. Walker L. Cisler
Retired Chairman of the Board
Detroit Edison Company

Mr. John T. Connor
Chairman of the Board
Allied Chemical Corporation

Mr. Richard C. Gerstenberg
Director, General Motors Corporation

Dr. H. Guyford Stever[1]
Former Director, National Science Foundation

Venezuela
Dr. José A. Mayobre
Banco Central de Venezuela

Ing. Ulises Ramírez
Executive Secretary
National Energy Council

[1] From October 1974 to October 1976.

ASSOCIATES

Canada
Mr. Marc LeClerc
Director General, Special Projects
National Energy Board

Dr. John Ralston Saul
Special Assistant to the Chairman
Petro-Canada

Finland
Mr. Seppo Hannus
Ministry of Trade and Industry
Energy Department

France
M. Bertrand Chateau
Institut Economique et Juridique de l'Energie
Centre National de la Recherche Scientifique

Dr. Maxime Kleinpeter
Eléctricité de France
Service Etudes Economique Générales

Germany
Dr. Georg Klotmann
Strategic Planning Department
Badische Anilin & Soda Fabrik AG

Klaus-Peter Messer
Rheinisch-Westfälisches Elektrizitätswerk AG

Dr. Dieter Schmitt
Head, Institute of Energy Economics
University of Cologne

Mr. Paul H. Suding
Economist, Institute of Energy Economics
University of Cologne

Italy
Dr. Oliviero Bernardini
Department of Technology Assessment
Research & Development Division
Montedison

Dr. Riccardo Galli
Director, Department of Technology Assessment
Research & Development Division
Montedison

Mr. William Mebane
Department of Technology Assessment
Research & Development Division
Montedison

Japan

Mr. Shinichiro Aoyama
Senior Staff Researcher
National Institute for Research Advancement

Mr. Kenichi Matsui
Senior Staff Economist
The Institute of Energy Economics

Mr. Yasuhiro Murota
Staff Economist
The Japan Economic Research Center

Mr. Mitsuo Takei
Director, Research Affairs
The Institute of Energy Economics

Mr. Hisashi Watanabe
Executive Director
Japan Alternative Energy Strategies Organization

Mexico

Ing. Gerardo Bazán
Assessor
Mexican National Energy Commission

Ing. Alberto Escofet
Assistant to the Executive Secretary
Mexican National Energy Commission

The Netherlands

Dr. André C. Sjoerdsma
Director, Future Shape of Technology Foundation

Netherlands/U.K.

Mr. Alan W. Clarke
Energy & Oil Economics Division
Shell International Petroleum Co., Ltd.
Royal Dutch/Shell Group

Dr. Hans DuMoulin
Head of Energy & Oil Economics Division
Shell International Petroleum Co., Ltd.
Royal Dutch/Shell Group

Dr. Gareth Price
Group Planning, Long Term Future
Shell International Petroleum Co., Ltd.
Royal Dutch/Shell Group

Norway

Mr. Henrik Ager-Hanssen
Deputy Managing Director
Statoil, Den Norske Stats Oljeselskap A/S

Mr. Kai Killerud
Manager, Power Plant Engineering
Scandpower A/S

Sweden

Dr. Harry Albinsson
Energy Secretary
Federation of Swedish Industries

Mr. Bertil Eneroth
Special Assignments
Skandinaviska Enskilda Banken

United Kingdom

Mr. Michael Clegg
Manager, Systems Group, Corporate Planning
British Petroleum Company, Ltd.

Dr. Edmund Crouch
Energy Research Group
Cavendish Laboratory
University of Cambridge

Dr. Richard J. Eden
Head, Energy Research Group
Cavendish Laboratory
University of Cambridge

Mr. Andrew R. Flower
Assistant Policy Analyst
Policy Review Unit
British Petroleum Company, Ltd.

United States

Mr. Walter F. Allaire
Director, Energy Resources
Allied Chemical Corporation

Mr. Steven Carhart[2]
Assistant Scientist
National Center for Analysis of Energy Systems
Brookhaven National Laboratory

---

[2] Consultant on Demand Studies, October 1974 to December 1975; Advisor on U.S. Studies from January 1976 to present.

Dr. Paul P. Craig[3]
Director, Energy & Resources Council
University of California

Dr. Henry L. Duncombe, Jr.
Chief Economist
General Motors Corporation

Ms. Sandra Fucigna[4]
Policy Analyst
National Science Foundation

Mr. Edward D. Griffith
Senior Consultant—Policy Analysis and Forecasting
Atlantic Richfield Company

Dr. Kenneth Hoffman[5]
Head, National Center for Analysis of Energy Systems
Brookhaven National Laboratory

Dr. H. Paul Root
Director of Economic Studies
General Motors Corporation

Dr. David Sternlight
Chief Economist
Atlantic Richfield Company

Venezuela

Dr. Felix Rossi-Guerrero
Minister Counselor for Petroleum Affairs
Venezuelan Embassy
Washington, D.C.

[3] Associate from October 1974 to April 1976.
[4] Associate from April 1976 to October 1976.
[5] Advisor on U.S. Studies from January 1976 to present.

STAFF

MIT Program Staff

Mr. Robert P. Greene
Program/Administrative Officer

Mr. Paul S. Basile
Program Officer

Mr. William F. Martin
Program Officer

MIT Support Staff

Ms. Elaine R. Goldberg

Ms. Susan M. Leland

Ms. Hedy Walsh

MIT Research Assistants

Mr. Richard Cheston

Mr. Miles Harbur

European Support Staff

Ms. Karin Berntsen, Norway

Ms. Adriana Cavagna, Italy

Ms. Dalia Jackbo, Norway

Ms. Margaret Martirosi, United Kingdom

ACKNOWLEDGMENTS

Institutions and Sponsors

Many organizations and institutions have provided direct financial support and other services for WAES. The Workshop gratefully acknowledges the following sponsors for their generous support of the Secretariat and related activities:

The Allied Chemical Foundation
The Atlantic Richfield Foundation
The Edna McConnell Clark Foundation
The General Motors Corporation
The Ford Foundation
The German Marshal Fund of the United States
The Andrew W. Mellon Foundation
The National Science Foundation
The Rockefeller Brothers Fund
The Rockefeller Foundation
The Alfred P. Sloan Foundation

The Workshop also gratefully acknowledges the many individuals and institutions in each country that have contributed financial or professional support for the Workshop or the WAES national studies, including:

Canada
National Energy Board
Petro-Canada

Denmark
Danish Research Council for Science
Niels Bohr Institute

Finland
Academy of Finland
Helsinki University of Technology
Ministry of Trade and Industry

France
Centre National de la Recherche Scientifique
Eléctricité de France
Société Générale

Germany
Badische Anilin & Soda Fabrik AG
Energiewirtschaftliches Institut, University of Cologne
Rheinisch-Westfälisches Elektrizitätswerk AG

Italy
Azienda Municipale Nettezza Urbana, Milano
Azienda Servizi Municipalizzati, Brescia
CISE
Confindustria
EGAM
ENI
FIAT
Istituto Economia Fonti di Energia (IEFE), Università L. Boccini, Milano
Istituto di Economia e Politica Industriale, Università di Bologna
Istituto di Fisica, Università di Napoli
Istituto Internazionale per le Ricerche Geotermiche, Pisa
IRI
Montedison
SNAM
Zanussi

Japan
Electric Machinery Industry Federation
Electric Power Industry Federation
Industrial Research Institute
Industrial Research Institute, Japan
Institute of Energy Economics

Institute for Future Technology
Japan Economic Research Center
Japan EXPO Fund
Japan Foreign Trade Council, Inc.
Japan Iron and Steel Federation
Japan Techno-Economic Society
Mitsubishi Research Institute
National Institute for Research Advancement
Nomura Research Institute
Petroleum Association of Japan

Mexico
Mexican National Energy Commission

The Netherlands
Akzo Zoutchemie
Centraal Planbureau
Estel Hoesch-Hoogovens
Ministerie van Economische Zaken
Ogem Holding
Paktank
Stichting Toekomstbeeld der Techniek
Verenigde Machinefabrieken

Netherlands/U.K.
Shell International Petroleum Company, Ltd.
Royal Dutch/Shell Group

Norway
Royal Norwegian Council for Scientific and Industrial Research
Elkem-Spigerverket A/S
Scandpower A/S
Statoil, Den norske stats oljeselskap A/S

A reference committee with observers from the Ministries and other institutions

Sweden
Gränges AB
Skandinaviska Enskilda Banken
Federation of Swedish Industries
Kraangede AB
National Board of Industry
Secretariat for Future Studies
Royal Academy of Engineering Sciences
State Power Board

United Kingdom
British Petroleum Company, Ltd.
Department of Energy
University of Cambridge:
—Churchill College
—Cavendish Laboratory
—Engineering Department

United States
Allied Chemical Corporation
Atlantic Richfield Company
Brookhaven National Laboratory
General Motors Corporation
Massachusetts Institute of Technology:
—Alfred P. Sloan School of Management
—MIT Energy Laboratory
National Science Foundation
Overseas Advisory Associates, Inc.

Venezuela
Ministry of Energy and Mines

This Report

A report of this scope would not have been possible without the assistance of many organizations, groups and individuals. I wish to thank the sponsors of the Workship Secretariat and the financial sponsors and technical supporters from each WAES country for their encouragement and support.

All Participants and Associates provided substantive input to each of the chapters in this report as they evolved through several drafts based on our studies and discussions. This Final Report of the Workshop reflects these many inputs.

Part I of this report, "Global Energy Futures," is the principal product of the Workshop. Michael Clegg prepared the first draft of Part I and Gerald Leach prepared the next two drafts. I gratefully acknowledge their contributions in organizing and expressing the complex technical, economic and political issues described. The final two drafts were extensively reviewed and revised by all Participants and Associates.

Part II of this report, "A Framework for the Energy Dialogue," is also the direct result of the work of the entire Workshop. It differs from Part I because one draftsman from the Associates or the MIT staff took principal responsibility for preparing the individual chapters.

Chapter 1, Mapping the Future: The WAES Approach, was written by Paul Basile, David Sternlight, and Gerald Leach.

Chapter 2, Energy Demand and Conservation, was written by Paul Basile.

Chapter 3, Oil, was written by Andrew Flower.

Chapter 4, Natural Gas, was written by William Martin and Andrew Flower.

Chapter 5, Coal, was written by Edward Griffith and Alan Clarke. A special study provided by the Bechtel Corporation added importantly to this chapter.

Chapter 6, Nuclear Energy, was written by Carroll Wilson; revisions were agreed on by Participants at the final meeting of the Workshop.

Chapter 7, Other Fossil Fuels and Renewables, was coordinated by William Martin with assistance from Sandra Fucigna, Edward Griffith, Marc LeClerc, John Page, John Saul and Leonard Topper.

Chapter 8, Critical Problems: Energy Demands and Supplies Mismatched, was written by Paul Basile, David Sternlight, and Richard Eden.

Appendix I, Energy and Economic Growth Prospects for the Developing Countries: 1960-2000, was written by William Martin and Frank J.P. Pinto (Consultant) of the Economic Analysis and Projection Department of the World Bank.

Each of the first eight chapters was extensively reviewed by the Associates. John Ravage provided editorial assistance. Bunji Tagawa and Diane Leonard-Senge prepared the visuals for the entire book from sketches developed during the course of the WAES work. Osgood Nichols was helpful in many aspects of the preparation of this report and Walter Tower had principal responsibility for the layout and design of this book. Robert Greene, manager of operations throughout the project, coordinated the production of the report through the various drafts with the able assistance of Paul Basile and William Martin. Susan Leland, my secretary since the beginning of the project, Elaine Goldberg, Roberta Ferland and Hedy Walsh typed the many drafts of the chapters and sent copies to all Workshop members—a process that involved handling hundreds of thousands of manuscript pages.

The Work Program of WAES

Our methods of work had to be adapted to the fact that members of WAES were located in 15 countries. Development of a common methodology for making projections of future energy supply and demand and discussion of the results of such studies made by WAES national teams required frequent meetings of Associates and periodic review and direction of the work by Participants.

Between October 1974 and the final meeting in February 1977 there were 13 meetings of Associates lasting 1-2 weeks and 7 meetings of Participants lasting several days. These meetings were held in various parts of the world and permitted discussions with individuals concerned with energy policy in several of the WAES countries.

Economy of travel and the convenience of all members influenced our choice of locations. WAES national teams were hosts for meetings in France, Germany, Italy, Japan, Mexico, Norway, Sweden, the United Kingdom and the United States.

Other Contributors

Dr. H. Guyford Stever, former Director of the National Science Foundation, was a Participant in the Workshop from its inception until October of 1976 when it became necessary for him to resign because of the demands of his new position as Science Advisor to the President of the United States. Dr. Paul P. Craig served as his Associate during the first year and a half of the Workshop and Ms. Sandra Fucigna served as his Associate for the last year.

Pierre Aigrain of France, Fritz Böttcher of The Netherlands, Senator Henry L. Jackson of the U.S.A. and Sir Ronald Prain of the United Kingdom were Advisors to me during the early stages of the project, and I gratefully acknowledge their guidance and encouragement. William T. Golden of New York first encouraged me to undertake this ambitious project.

Other contributors were:

Dr. Gerald Leach
International Institute for
Environmental Development
London, England

Mr. Osgood Nichols
Osgood Nichols Associates, Inc.
New York, New York

Prof. John K. Page
Sheffield University
Sheffield, England

Mr. Frank J.P. Pinto
Consultant
Economic Analysis and Projections
World Bank
Washington, D.C.

Mr. John Ravage
Editor
Northfield Mt. Hermon School
Greenfield, Massachusetts

Ms. Diane Leonard-Senge
Graphic Artist
Cambridge, Massachusetts

Mr. Bunji Tagawa
Graphic Designer
New York, New York

Mr. Walter T. Tower
President, Nimrod Press
Boston, Massachusetts

On behalf of the members of the Workshop, I wish to express our great appreciation to all who have helped to make this report possible.

CARROLL L. WILSON

INTRODUCTION

After two and a half years of collaborative study and analysis of energy prospects to the year 2000, the Workshop on Alternative Energy Strategies presents this summary of its findings and conclusions.

The task we took on—global* energy assessment to the year 2000—was enormous in relation to our time and resources. It is the first such study. We could not study all issues in equal depth, so we had to concentrate on those which seemed most important in revealing prospective changes in the energy situation over the next 25 years. We also believed that we should publish our principal findings as soon as possible despite the risk of uneven quality in our analysis.

We concentrated on estimates of demand and potential supplies of the principal fuels which would be important during the next 25 years—oil, gas, coal and nuclear—as well as prospective gaps between desires for such fuels and probable supplies. We considered some of the issues arising from such prospective energy gaps for national policy and for international action.

We were well aware that the implications of our findings could have great importance beyond the 20th century. All fossil fuel resources, which might be considered a form of stored solar energy, will eventually be depleted in meeting man's enormous appetite for energy.

Such perceptions lead one naturally to consider the prospects for solar energy and ways of collecting, storing and using it. All renewable sources must play a growing role in the next century. Given their future importance, our treatment of these subjects is inadequate for two reasons. First, we do not expect them to become significant energy sources before 2000. Second, because these technologies are only now emerging, it is difficult to estimate their costs reliably or fully understand their likely effects on the environment. We hope that others will give these resources closer attention than we could.

Our demand projections to year 2000 assume much energy conservation in the scenarios of each WAES national team. We think these projections span a plausible range of global needs. They rest on

* The "world" we studied is the World Outside Communist Areas (WOCA).

assumptions of economic growth rates, energy prices, and vigor of national policy that seem to us consistent with reasonable expectations of the range of real economic growth for the future.

The basic goals of our study have been a) to develop a useful method of projecting national supply and demand for energy, b) to study supply and demand to 1985 and 2000 using this method for the 13 WAES countries that together consume most of the world's energy, c) to develop methods for estimating global production of oil, gas, coal and nuclear, and d) to determine whether prospective global shortages of certain fuels are likely to occur, when, and how rapidly they might grow.

Our supply studies, especially our analysis of the range of future world oil production, may be our most important contribution to the study of energy futures. Our evaluation of the other principal fuels—natural gas, coal and nuclear—has two aims: to show the potential of such fuels to fill a prospective gap created by a growing shortage of oil, and to identify the obstacles to making the transition from oil to other fuels in time to meet the world's energy needs.

We have assumed that substantial improvements in energy efficiency will decrease the growth rate of energy demand and that supplies of most fuels will increase because of vigorous national efforts. Still, we are left with large prospective shortages. If projected demand is to be met, such shortages will have to be filled by alternatives such as other fossil fuels and renewables whose technical and economic feasibility and rates of implementation we understand poorly.

We ask our readers' patience in adjusting to our basic unit for measuring the rate of energy use—one million barrels of oil per day (MBD), or its energy equivalent in other fuels—one million barrels per day of oil equivalent (MBDOE). One MBD of oil is approximately equivalent to 50 million tons per year of oil. After much debate we chose this unit for several reasons—a central reason is that we foresee a transition from dependence on oil, and oil flow is usually measured in barrels per day. Also, in a world projected to need 160-210 MBDOE in the year 2000, replacement energy systems must be very large—millions of barrels a day of oil equivalent—to replace oil.

Naturally, to carry out parts of our work we have depended upon individual expertise among us in many areas such as traditional fossil fuels, nuclear energy, alternative fuels, environmental and climate change, energy-intensive industries, energy conservation, economics, etc. But our common task has been to understand the analysis, to reach agreement on its validity, and to reach a consensus on its principal implications.

This is the final report of the Workshop. The Workshop has also prepared three technical reports: *Energy Demand Studies: Major Consuming Countries* (MIT Press, November 1976); *Energy Supply to the Year 2000: Global and National Studies*; and *Energy Supply-Demand Integrations to the Year 2000: Global and National Studies* (MIT Press, June 1977).

This report, *ENERGY: Global Prospects 1985-2000*, represents the collective efforts of the Workshop members based on these technical studies and our deliberations together. It is divided into two parts: Part I—Global Energy Futures, and Part II—A Framework for the Energy Dialogue. The differences in content and style between Parts I and II reflect the differences in the way they were prepared. Part I is largely the result of a group effort, while the chapters in Part II were primarily the responsibility of one individual.

Part I, Global Energy Futures, is the principal product of the Workshop Participants; it provides an overview of the total energy scene, its problems and some possible solutions. This part of the report has been reviewed and discussed extensively by all Participants prior to finalization.

Part II, Framework for the Energy Dialogue, contains the detailed assessments of individual facets of the energy scene through the year 2000.

Participants and Associates have taken part in the study as individuals and this report therefore does not necessarily represent or reflect the views of any public or private organizations with which WAES members may be associated.

No single member had either the time or the expertise to judge every topic covered; moreover, on a few subjects individual members have held differing views. The report in such cases has sought the consensus of the majority, and it is accepted that each member does not necessarily subscribe to all of the statements in the report.

Nevertheless all Workshop members agree on the general analysis and the main findings of the report, and all believe that this first global assessment of the energy situation to the year 2000 carries an urgent and important message.

CLW

PART I

GLOBAL ENERGY FUTURES

PART I: GLOBAL ENERGY FUTURES

Conclusions — Mapping the Future — The Critical Problem: Gaps Between Supply and Demand — The Decline of the Oil Era — Energy Demand — Conservation — Coal — Natural Gas — Nuclear Energy — Hydroelectricity — Other Fossil Fuels — Geothermal — Energy Pricing — Environment and Climate — Uncertainties — National and International Issues

Conclusions

After two years of study we conclude that world oil production is likely to level off—perhaps as early as 1985—and that alternative fuels will have to meet growing energy demand. Large investments and long lead times are required to produce these fuels on a scale large enough to fill the prospective shortage of oil, the fuel that now furnishes most of the world's energy. The task for the world* will be to manage a transition from dependence on oil to greater reliance on other fossil fuels, nuclear energy and, later, renewable energy systems.

Our major conclusions are as follows:

1. The supply of oil will fail to meet increasing demand before the year 2000, most probably between 1985 and 1995, even if en-

---

* This study of energy has focused on the World Outside Communist Areas (shortened in this report to the acronym WOCA) which is often called the non-Communist world. The U.S.S.R. and China are major energy producers and suppliers, although their trade in fuels with countries in WOCA has to date been relatively small. In the WAES projections we have assumed that this situation would continue to the end of the century.

ergy prices rise 50% above current levels in real terms. Additional constraints on oil production will hasten this shortage, thereby reducing the time available for action on alternatives.

2. Demand for energy will continue to grow even if governments adopt vigorous policies to conserve energy. This growth must increasingly be satisfied by energy resources other than oil, which will be progressively reserved for uses that only oil can satisfy.

3. The continued growth of energy demand requires that energy resources be developed with the utmost vigor. The change from a world economy dominated by oil must start *now*. The alternatives require 5 to 15 years to develop, and the need for replacement fuels will increase rapidly as the last decade of the century is approached.

4. Electricity from nuclear power is capable of making an important contribution to the global energy supply although worldwide acceptance of it on a sufficiently large scale has yet to be established. Fusion power will not be significant before the year 2000.

5. Coal has the potential to contribute substantially to future energy supplies. Coal reserves are abundant, but taking advantage of them requires an active program of development by both producers and consumers.

6. Natural gas reserves are large enough to meet projected demand provided the incentives are sufficient to encourage the development of extensive and costly intercontinental gas transportation systems.

7. Although the resource base of other fossil fuels such as oil sands, heavy oil, and oil shale is very large, they are likely to supply only small amounts of energy before the year 2000.

8. Other than hydroelectric power, renewable resources of energy —e.g., solar, wind-power, wave-power—are unlikely to contribute significant quantities of additional energy during this century at the global level, although they could be of importance in particular areas. They are likely to become increasingly important in the 21st century.

9. Energy efficiency improvements, beyond the substantial energy conservation assumptions already built into our analysis, can further reduce energy demand and narrow the prospective gaps between energy demand and supply. Policies for achieving energy conservation should continue to be key elements of all future energy strategies.

10. The critical interdependence of nations in the energy field requires an unprecedented degree of international collaboration in the future. In addition it requires the will to mobilize finance, labor, research and ingenuity with a common purpose never before attained in time of peace; and it requires it now.

Failure to recognize the importance and validity of these findings and to take appropriate and timely action will almost certainly result in a world different from the one on which these projections have been based. Failure to act could lead to substantially higher energy prices as the supply/demand imbalance becomes more apparent—with the depressant effects on the economies of the world and the consequent frustration of the aspirations of the less-developed countries. The major political and social difficulties that might arise could cause energy to become a focus for confrontation and conflict.

In addition, the longer the world delays facing this issue the more serious the outcome will be. Even with prompt action the margin between success and failure in the 1985-2000 period is slim. Time has become one of the most precious of our resources. Recognizing the importance of time and the need to respond can help us through the period of transition that lies ahead.

Part I is a summary of the analysis on which these conclusions are based. Part II is a more detailed analysis of demand and conservation, oil, gas, coal, nuclear and other fuels, and finally the system we used to combine our estimates of future supply and demand in order to see where and when prospective shortages might appear. Much greater detail will be found in three Technical Volumes.*

---

* *Energy Demand Studies: Major Consuming Countries* (November 1976, MIT Press), *Energy Supply to the Year 2000: Global and National Studies*, and *Energy Supply-Demand Integrations to the Year 2000: Global and National Studies* (June 1977, MIT Press).

Mapping the Future

The Workshop on Alternative Energy Strategies (WAES) was formed to examine these issues. We have tried to provide a clear and consistent framework for national debate and international dialogue on longer term energy strategies. By clarifying the kinds of energy problems that lie ahead, one can draw out the main implications that nations acting alone and together must consider.

To do this we had to first examine closely the 1985-2000 period when the critical energy problems might appear and, second, consider national and international issues in an integrated way. No nation is free to act alone—yet each nation must make decisions. Our method of projecting energy futures recognizes this basic fact. We begin by defining plausible future states of the World Outside Communist Areas (WOCA) using assumptions about economic growth, energy prices and national policies. We then consider in

Figure I-1 The WAES Scenario Framework

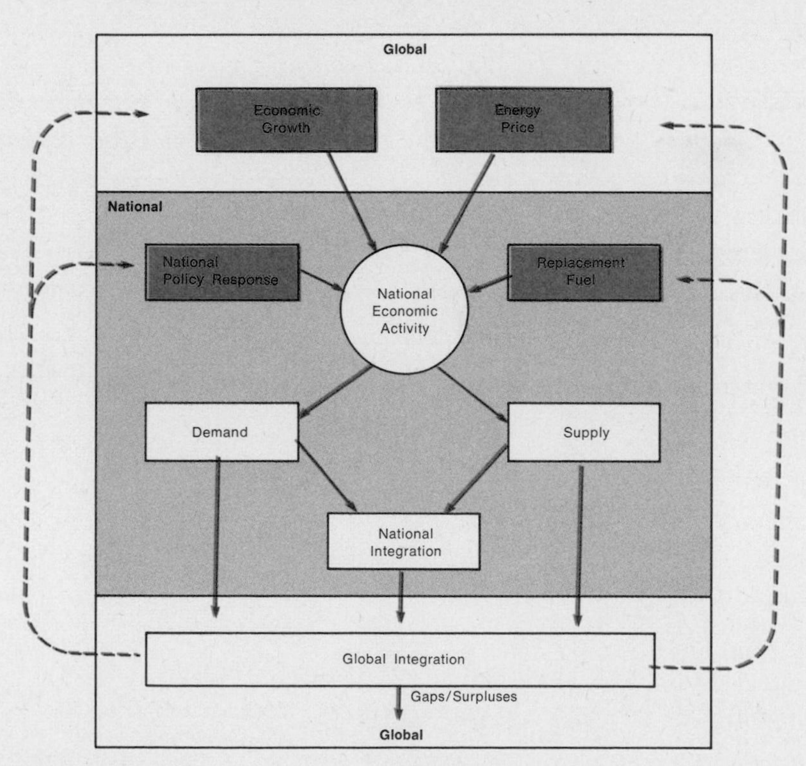

Figure I-2 The WAES Scenarios

	Factors That Influence Future Economy	Variables		
1977-1985	World economic growth rate*: high (6%) or low (3.5%)	High		Low
	Oil price: rising ($17.25) constant ($11.50) or falling ($7.66)	17.25	11.50	7.66
	National policy response: vigorous or restrained	Vig		Res
1985-2000	World economic growth rate: high (5%) or low (3%)	High		Low
	Energy price: rising ($17.25) or constant ($11.50)	17.25	11.50	
	Gross additions to oil reserves: 20 BB/YR or 10 BB/YR	20		10
	OPEC oil ceiling: 45 MBD or 40 MBD	45		40
	Principal replacement fuel: coal or nuclear	Coal		Nuc

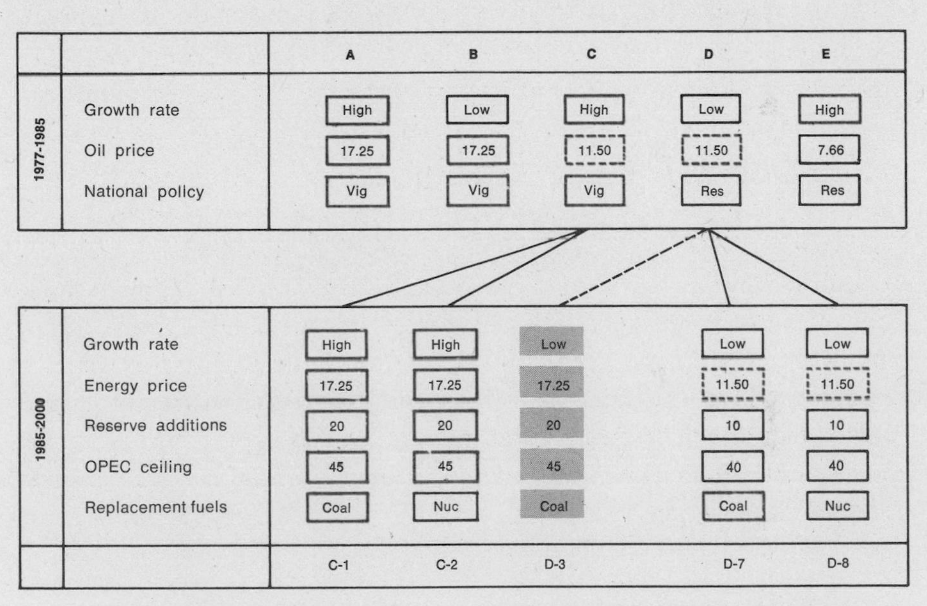

* The period from 1973 to the end of 1975 is assumed to correspond to actual world economic conditions, with recovery to 1973 levels by the end of 1976 postulated.

National economic studies done under these assumptions, when summed, actually result in global rates of 5.2 and 3.4% to 1985 and 4.0 and 2.8% from 1985 to 2000.

We assume that the scenario variables are approximately independent of each other, within a certain range of values. Any combination of values *may*, then, be possible and no combination is automatically excluded from consideration.

detail how each nation or region might respond within this global environment. Finally, we examine how these individual responses combine to produce a new global environment in which nations must formulate their goals and policies for energy. This framework is outlined in Figure I-1, and described in more depth in Chapter 1. Details on the cases that we studied are shown in Figure I-2. Throughout this report, we refer to these scenario cases by letter and number (e.g., Case C-1, Case D-8, etc.).

Within this framework we examined, by nation and by region, the desired demands for fuels and, separately, the maximum potential capacity to supply them. These demand and supply studies are outlined in this chapter and discussed in detail in Chapters 2 through 7. They are important in their own right, but combining them reveals the most important features of the future. Our technique for studying supply-demand integration (Chapter 8)—that is, whether supply can balance demand—exposes potential fuel gaps or surpluses. It reveals any mismatches between the amount of fuel each nation would like to use (its *desired demand*), assuming it were available under the explicit scenario assumptions, and the amount of fuel it could potentially produce (its *maximum supply*) or import under the same assumptions.

These prospective *energy gaps* or *energy surpluses* are the crux of our findings. In the real world they cannot actually occur, since total fuel supplies and consumption, imports and exports, must always balance: no group of nations can import more oil than producer countries have available for export. Adjustments to avoid gaps would occur, perhaps through different prices or economic growth from those assumed in our scenarios or through anticipatory government policies. Hence the existence of gaps in our projections really illuminates the degree of change that must and will occur. Such changes must, however, occur in an ordered, controlled fashion if sharp and sudden adjustments that could amount to an "energy crisis" are to be avoided.

The Critical Problem: Gaps Between Supply and Demand

When we look ahead over the remainder of the century, our scenarios paint an increasingly disconcerting picture after 1985. If we consider first the most critical fuel—oil—we find that potential (maximum) supplies could be sufficient to meet desired demands up to 1985. But thereafter, a gap between supply and demand appears and widens rapidly, giving large deficits in oil supply in all our scenarios by 2000. This prospective gap can be measured as the shortfall between oil imports desired by the major consuming countries and the maximum exports available from oil producing and exporting countries. Table I-1 shows how this import-export gap equals from one-quarter to one-third of expected ("desired") oil imports in all of the WAES scenarios for 2000.

Table I-1 Summary of Oil Balance in the Year 2000

Economic Growth:	High	High	Low	Low
Energy Price (1985-2000):	Rising	Rising	Constant	Constant
Principal Replacement Fuel:	Coal	Nuclear	Coal	Nuclear
WAES Scenario Case:	*C-1*	*C-2*	*D-7*	*D-8*
Major Importer's Desired Imports		(all numbers are in MBD)		
North America*	10.4	10.7	15.8	15.8
Western Europe	16.5	16.4	13.2	12.5
Japan	15.2	14.4	8.2	7.9
Non-OPEC Rest of WOCA	11.2	9.5	9.6	9.0
International Bunkers**	5.4	5.4	4.5	4.5
Total Desired Imports	58.7	56.4	51.3	49.7
Major Exporters' Potential Exports				
OPEC***	38.7	37.2	35.2	34.5
Prospective (shortage) or surplus	(20.0)	(19.2)	(16.1)	(15.2)
as a percentage of total WOCA potential oil production	27%	26%	28%	26%

* Takes account of domestic production of oil shale and oil sands in addition to conventional oil production.

** International bunkers represent the oil used in international shipping.

*** OPEC potential exports equal OPEC potential production minus OPEC internal demand.

North America: Canada, U.S.A.

Western Europe: *WAES-Europe* plus *non-WAES Europe*

WAES-Europe		*non-WAES Europe*	
Denmark	The Netherlands	Austria	Luxembourg
Finland	Norway	Belgium	Portugal
France	Sweden	Greece	Spain
F.R.G.	U.K.	Iceland	Switzerland
Italy		Ireland	

OPEC (Organization of Petroleum Exporting Countries): Algeria, Ecuador, Gabon, Indonesia, Iran, Iraq, Kuwait, Libya, Nigeria, Qatar, Saudi Arabia, United Arab Emirates (Abu Dhabi, Dubai and Sharjah), Venezuela.

Non-OPEC Rest of WOCA: all other countries outside Communist areas.

WOCA: World Outside Communist Areas.

Natural gas poses a different set of problems. Exploitable reserves in WOCA are large, but they are far from the major consuming countries. And in some of the major producing countries, where the infrastructure already exists, reserves are likely to decline within the next few years. Thus, the main constraint on supply is likely to be not the level of reserves, but the problem of moving sup-

ENERGY UNITS

In this report we have generally expressed the rate of energy use (both for fuels and electricity delivered to consumers and for supply and demand of primary energy) in terms of *millions of barrels per day of oil equivalent* (MBDOE). This measure is based upon the conventional unit of a "barrel of oil equivalent" with a gross calorific value of 5.8 million British thermal units (Btu's).

In many places throughout the text we have also employed other units of measurement, for example exajoules (10^{18} joules), quadrillion Btu (10^{15} Btu or "Quads") or units applicable to particular energy sources, such as tons of coal or cubic meters of gas.

The calorific value of the various types and sources of coal, crude oil and natural gas varies widely. Thus, the following equivalents to our basic energy unit are approximate.

1 MBDOE = 50 million tons of oil equivalent (TOE) per year
= 76 million metric tons of coal equivalent (TCE) per year
= 57 billion (10^9) cubic meters of natural gas per year

plies from producers to consumers. Natural gas demands in 2000 can be met if increasing volumes of liquefied natural gas (LNG) are imported from OPEC countries, or if large amounts of synthetic gas are manufactured.

In the case of coal, we considered the supply levels that may be available and the contribution they might make to a reduction in the projected oil gap. But consumer preference for fuels other than coal results in surpluses—excesses of potential coal supply over desired coal demand. Such coal surpluses could be exported by some countries and imported by others to try to fill the oil gap.

Nuclear energy can contribute considerably to energy supply toward the end of the century. In the WAES projections, electricity from nuclear energy represents one of the important substitutes for

$= 2.2$ exajoules (10^{18} joules) per year
$= 530 \times 10^{12}$ kilocalories per year
$= 2.1 \times 10^{15}$ Btu (Quads) per year
$= 620$ terawatthours (10^9 kWh) per year

Additional equivalents are given in Chapter 1.

To produce an electrical output of 620×10^9 kWh per year (1 MBDOE) would require power stations of 100 GWe (= 100,000 megawatts) installed capacity, given an average load factor of 70%. If the average efficiency of the power stations is 35%, one needs a fuel input of $\frac{1}{0.35}$ (= approximately 2.8) MBDOE. The difference between the input of energy as fuel and the output of energy as electricity (in this example, 1.8 MBDOE) is the transformation loss in electricity generation. We include this loss under the heading "processing losses," along with the losses which occur in all other energy conversion processes.

In our discussion of primary energy supply and demand we have adopted the convention of expressing electricity from primary sources (nuclear, hydro, geothermal, etc.) in terms of the fuel input (generally expressed in MBDOE) that would be required to produce the equivalent amount of electricity output in fossil-fueled power stations.

oil and gas in several demand sectors. This substitution, however, is conditional on a resolution of the uncertainties which at present impede its development. In our low and high nuclear cases for 2000 (Cases D-7 and C-2) nuclear energy is expected to supply from 14 to 21% of primary energy requirements.

Figure I-3 brings together many of these estimates of prospective gaps and surpluses for the WAES scenarios for 1985 and 2000, with the 1972 figures as a baseline comparison. More detailed findings and supporting discussion are contained in Chapter 8.

In 1985 (see Figure I-4), the C and D cases with oil price constant in real terms show a well-balanced picture. However, in Cases A and B (rising oil price to 1985), there is projected to be a potential *surplus* of certain fuels over the desired demands for them.

The implication is that the price of oil (rising to $17.25* per barrel by 1985 in constant 1975 U.S. dollars) assumed in these cases might be higher than needed to bring on sufficient alternative supplies and encourage energy conservation.

Of the 1985 scenarios, Case E, in which we assumed high economic growth, falling oil price and restrained national policy response, produces awkward results: there are prospective shortfalls in certain fuels and in total energy. The price of oil, which falls to $7.66 per barrel by 1985, is too low to bring on enough supplies to meet the high energy demand levels resulting from the combination of high economic growth, restrained policy and falling price. To bring supplies and demands into balance, policies could force major fuel substitutions—such as getting industrial consumers to use more coal and less oil—or there could be natural adjustments toward higher prices, or lower economic growth, or policies to encourage greater energy conservation and more vigorous national supply expansion.

The year 2000 scenarios present much more challenging problems. In most of our projections, there are prospective energy gaps—characterized by large shortfalls of oil and surpluses of coal. The main findings of our *unconstrained* integrations of supply and demand for a selected case are illustrated in Figure I-3, which shows our projections of the desired demands for and potential supplies of each fuel. Notice the growing oil deficit and coal surplus by 2000. The required growth in natural gas production and in the supply of nuclear, hydro, and "other" energy sources is also shown in Figure I-3.

The fact that *total* energy supply and demand in WOCA generally differ only by a small percentage (between 5% and 8%) in our scenarios to 2000 may lead people to believe that very small adjustments could produce a balance between supply and demand. The underlying situation is really much more serious. Much larger imbalances are projected between the supply of and demand for such preferred fuels as oil, as shown in Figure I-3. The prospective shortfall of oil in 2000 is between 26 and 28% of the total potential oil production in our cases. These prospective fuel gaps must somehow be eliminated. How might this be done?

Using the integration procedure described in Chapter 1, we

---

* All oil or energy prices are given in constant 1975 U.S. dollars per barrel of Arabian light crude oil, f.o.b. Persian Gulf.

Figure I-3 Energy Demand and Supply in the World Outside Communist Areas

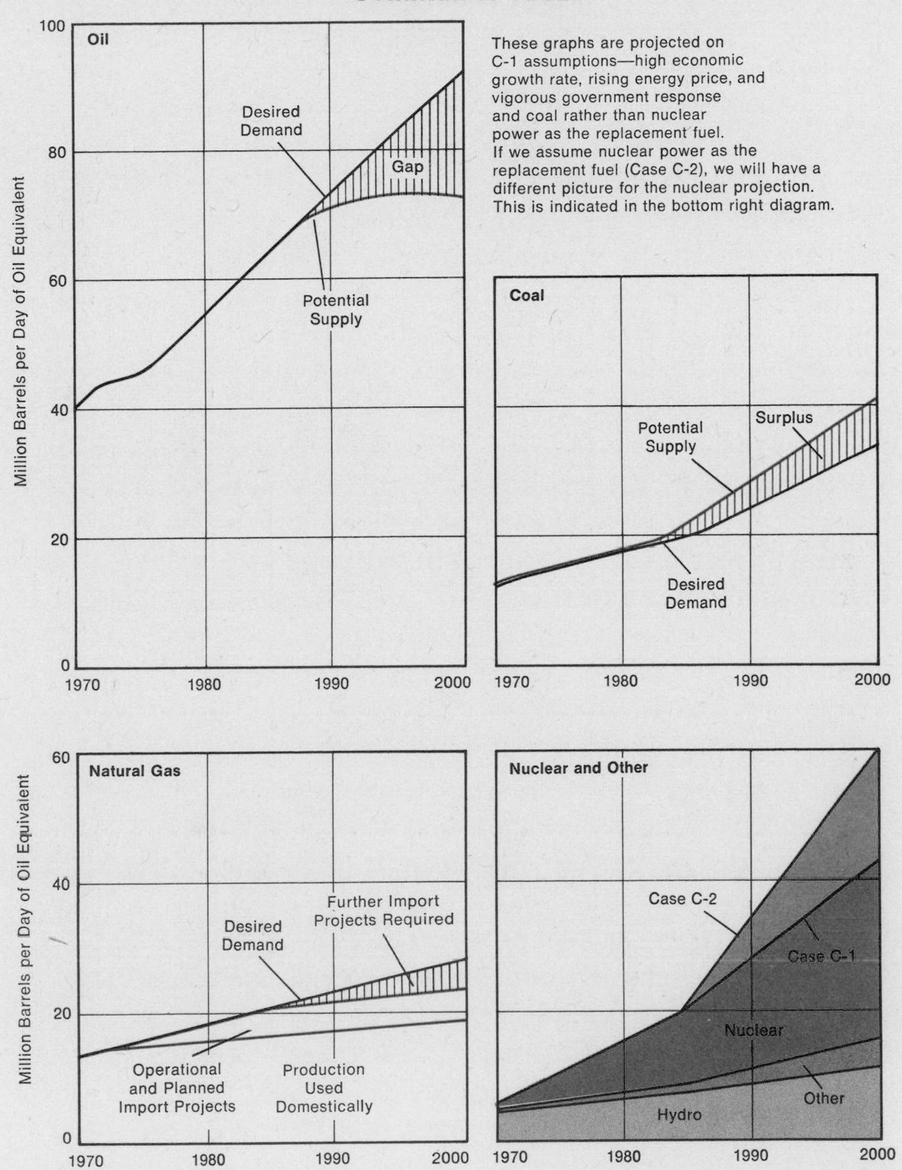

asked if there was any possible allocation of available fuels to consumers which would allow all energy demands to be met. The type of fuel mix foreseen in these *constrained* scenarios is very different from that existing today. To balance supply and demand while maintaining economic growth at the WAES case prices, use of oil and

Figure I-4 Energy Supply and Demand, WOCA, 1985

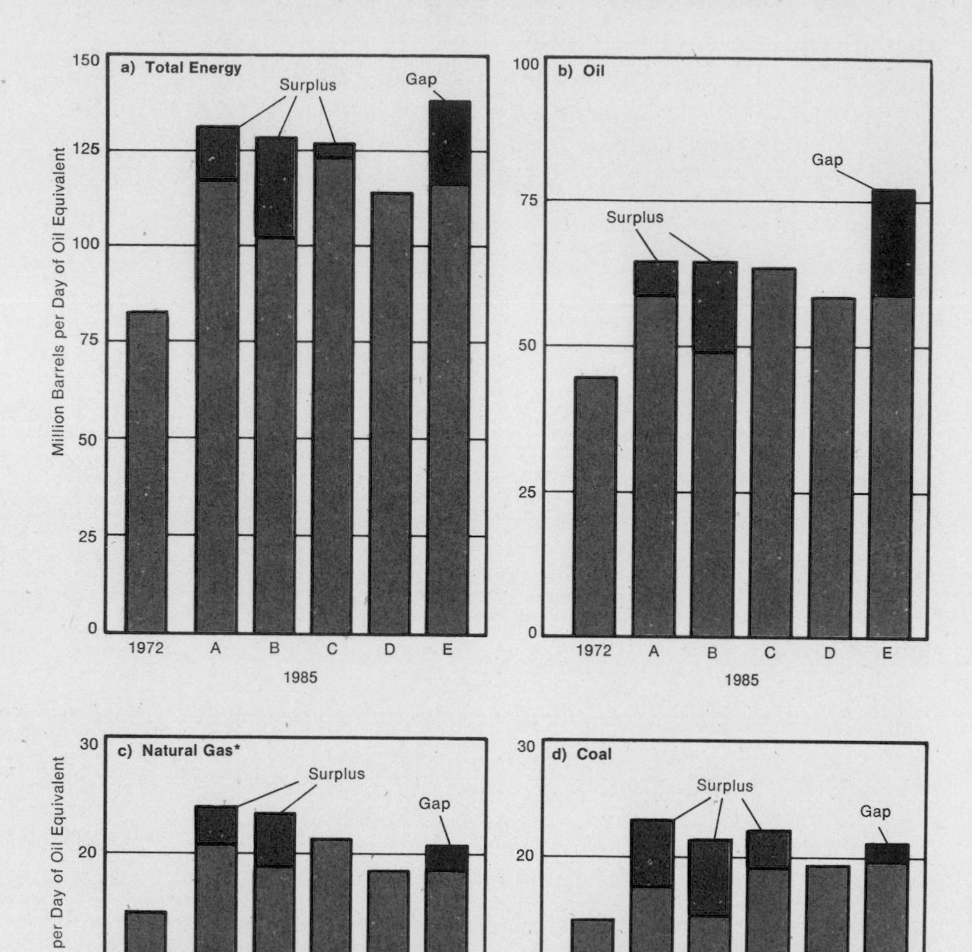

* Natural gas supply figures are *net* production; gross production figures would be slightly higher due to processing losses in gas transport.

natural gas in electricity generation in such constrained cases must become negligible (about 10-15%) in most regions of the world as compared with an average of 35% in 1972—the balance to be made up by a massive shift to nuclear and coal. In addition, scarce oil is shifted from the domestic and industrial sectors in order to reserve it for use almost exclusively in transport and as petrochemical feedstocks. Vigorous policies to influence the pattern of fuel use would be required to effect these changes.

The four main constrained scenarios examined to 2000 also show that even if fuel substitutions were taken to their limits—even if prospective surpluses of coal were used to fill (in part) prospective shortages of oil—total energy gaps would be substantially reduced, but not eliminated. The failure to meet the energy demands projected on the basis of our scenarios suggests that the plausible futures we postulated may not be attainable. Thus some or all of the following might occur:

a) Economic growth may not achieve a level of even 3% per year (the low WAES assumption) throughout the 1975-2000 period. This could have important repercussions such as high unemployment in many countries. But the largest effects may be on the developing countries because of their dependence on trade with the industrialized regions of the world and their inability, without aid, to pay for essential imports.

b) Government responses may be different from the vigorous ones already assumed—for example more vigorous actions to encourage conservation.

c) Energy prices may move up beyond the range assumed. So long as these increases are slow, steady, and predictable, they might allow adequate time for adjustments which will encourage energy savings and stimulate alternative fuel supplies. Such gradual adjustments would be of benefit in the 1985-2000 period as well as during the longer-term transition to new and sustainable patterns of energy supply and use that must be developed for the 21st century. Alternatively, rapidly rising energy prices could increase inflation, exacerbate balance of payment problems, produce low or negative economic growth rates, and increase unemployment.

WAES has not analyzed these particular price and growth implications in detail. However, we tentatively examined an alternative scenario (Case D-3 of Figure I-2) which combined the lower economic growth and higher energy price assumptions of our other year

2000 cases. We did not analyze this case in full detail, but we found clear indications that overall energy supply and demand would "balance," that there would be no prospective gaps in any fuel, and that the peak in oil supply would not occur until after the end of the century. With lower additions to oil reserves, the resulting prospective shortfall in oil supply could occur before 2000 but would be much smaller than in the other cases. Thus there are scenarios in which the need for massive adjustments might be delayed. But it is only a delay, not a solution.

The transition in energy use to the end of the century may be eased if other alternatives such as more intensive conservation, major new initiatives on renewables, and the development of oil shale, heavy oils and oil sands were vigorously pursued.

Another approach to "closing the gaps" is to ask how each nation might respond to such prospective global gaps. Beyond market adjustments in energy prices and economic growth, efforts would probably be intensified to reduce energy demand, and to substitute alternative fuels such as coal, nuclear, other fossil fuels and solar for oil and gas, the most limited fuels. Because such efforts would deeply involve the internal policies and consequences of individual nations, at this stage they go beyond the scope of an international project such as WAES. We have limited our analysis to showing what would happen if each nation made its energy policies *without* regard to the global results 15 to 25 years ahead.* Others will have to consider the implications of our projections, weigh alternative choices, and see what these mean in international terms for future world energy demand and supply.

The basic danger of the world energy situation is that it could become critical before it seems serious. Most governments and businesses—for many legitimate reasons—focus their efforts within a time horizon of 5 to 10 years. With such relative shortsight, the energy future does not seem serious. Although there are some notable exceptions to this, for example a recent OECD** report, the demand and supply projections made by many groups appear to balance and

* National unconstrained studies of this kind can be found in *Energy Supply-Demand Integrations to the Year 2000: Global and National Studies* (MIT Press, 1977).

** OECD is the Organization for Economic Cooperation and Development. The report cited is *World Energy Outlook*, published in Paris, 1977.

the gaps that open beyond 1985 are invisible or, if perceived, are often turned aside with the explanation that "something will turn up." WAES cannot identify such hoped-for somethings in the future. What we find is a range of opportunities for closing the gaps that all require enormous efforts in planning, intensive engineering efforts and major capital investment, with lead times usually of 10 or more years. And most of these efforts should be well under way by 1980-1985, which means starting them now. There is no time for procrastination.

The Decline of the Oil Era

Behind all the prospective energy gaps and imbalances that appear beyond 1985 is the inescapable fact that the time when the production of oil will plateau and then decline is clearly in sight.

It is hard to overdramatize the importance of oil as a world

Figure I-5 Energy Production and Consumption of the World (Including Communist Areas)

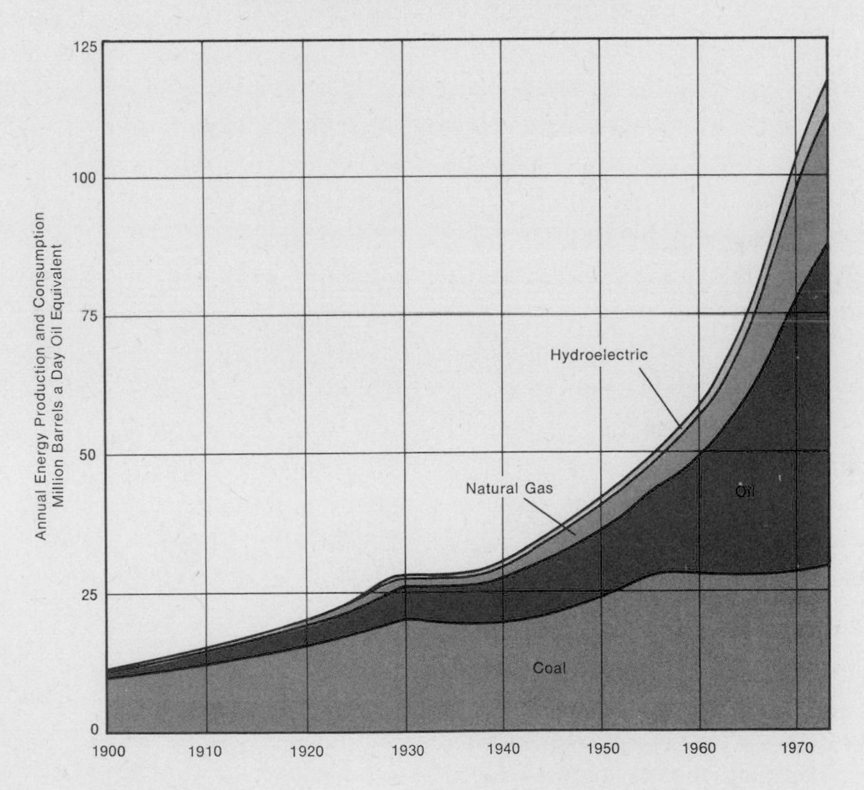

energy source. Figure I-5 shows how world energy production has increased during this century; almost the entire growth has been from oil, with the curve soaring upwards in the 1950's and 1960's due to the growth in world economic activity, the declining price of oil in real terms, and increasing environmental concerns with other fuels. Figure I-6 shows the world trade in oil in 1973. An immense and intricate system moves about 45 million barrels of oil every day from producer to consumer. In order to meet demand growth expectations it will have to be expanded to handle as much as 70 MBD by the late 1980's.

We looked at how world oil demand might grow given the WAES assumptions on economic growth and energy prices and found that future growth rates could fall significantly below the 6% per annum that prevailed in the decade before 1973. Our analyses showed annual growth rates for oil demand falling to the range 2.6 to 3.4% in the 1975-1985 period and to 1.5 to 2.6% from 1985-2000.

Next we looked at oil resources and potential supply. Estimates of total ultimately recoverable oil reserves are agreed by many experts to be of the order of 2000 billion barrels of which perhaps 75%-80% might be in WOCA. We took this as a central estimate

Figure I-6 The Flow of Oil in 1973

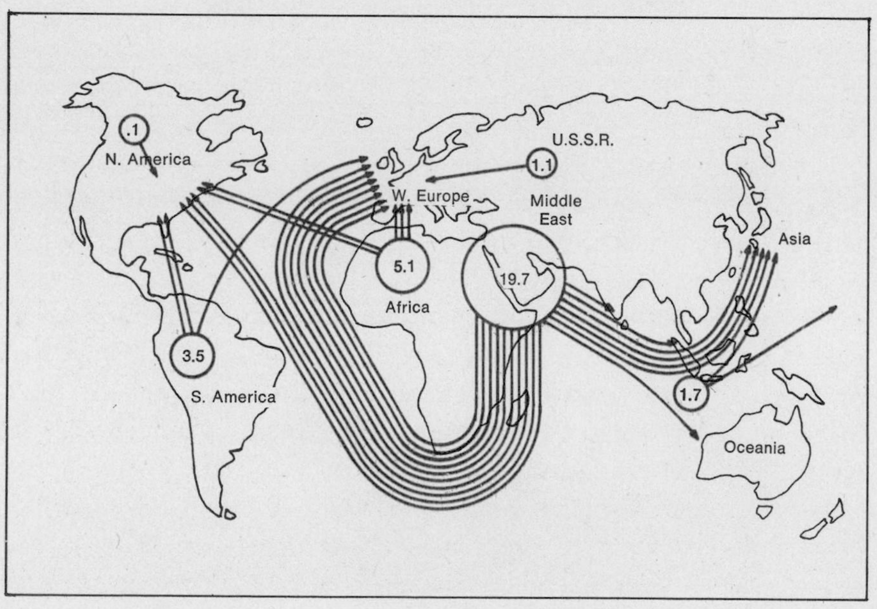

but we also considered the possibility of higher and lower reserve figures.

However, more important than the assessment of total potential reserves is the rate at which currently proven oil reserves are augmented by new discoveries, extensions to known fields, and better recovery techniques in the period to 2000. We made two assumptions for the rate of gross additions to reserves—a "high" rate of 20 billion barrels per year and a "low" rate of 10 billion barrels per year —compared with an historical rate of 18 billion barrels per year (based on backdating field extensions to the year in which a field was discovered, a procedure discussed in detail in Chapter 3). Taking into account those assumptions, we then developed an oil production profile based on oil production meeting oil demand until a technical production limitation—the reserve/production ratio or R/P ratio of 15/1 was encountered. We also investigated the potential productive capacity and possible levels of government-imposed ceilings for the principal OPEC producers. The results of this analysis of demand and supply show how world oil production would peak and decline—given our explicit assumptions.

These results are illustrated in Figure I-7 for high gross annual additions to oil reserves and high demand growth (Case C-1) and in Figure I-8 for low gross additions to oil reserves and low demand growth (Case D-8). In all our projections oil continues to play an important role in energy supply. Yet in every case a production peak or plateau occurs before the end of the century—in some cases well before.

The figure for gross additions to oil reserves of 20 billion barrels per year includes an allowance for enhanced recovery. Such improved recovery is potentially one of the ways in which large additions to oil reserves could be made. However, major improvements in recovery techniques will require extensive and costly research, development and demonstration. The main contribution of such improved recovery will most likely be a maintenance of an oil plateau for a longer period into the 21st century, rather than an increase in the level of peak oil production.

The fact is that at some future date the available supply of oil is unlikely to meet desired demand and such demands will have to be modified. Many national forecasts continue to project that large gaps between energy demand and indigenous supply beyond 1990

Figure I-7 Oil Production, WOCA (Case C-1)

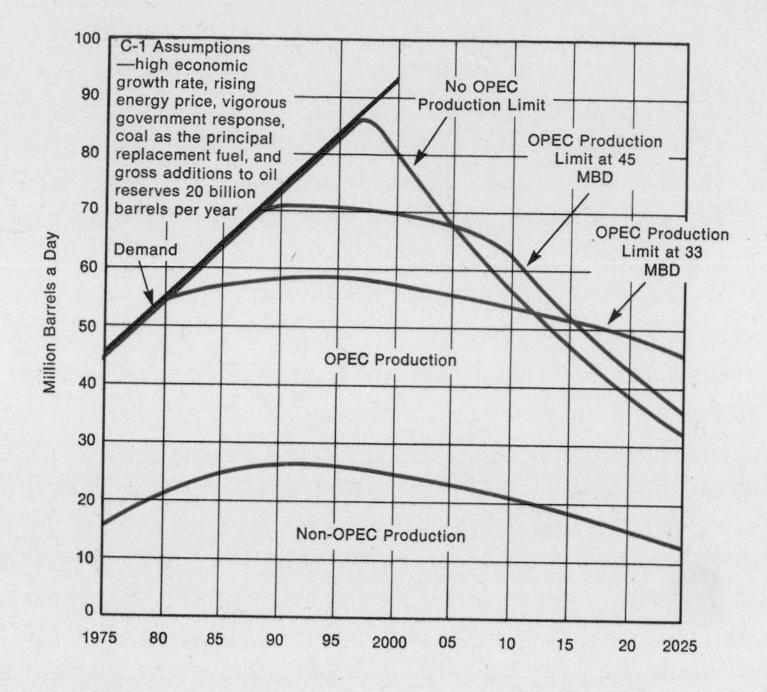

Figure I-8 Oil Production, WOCA (Case D-8)

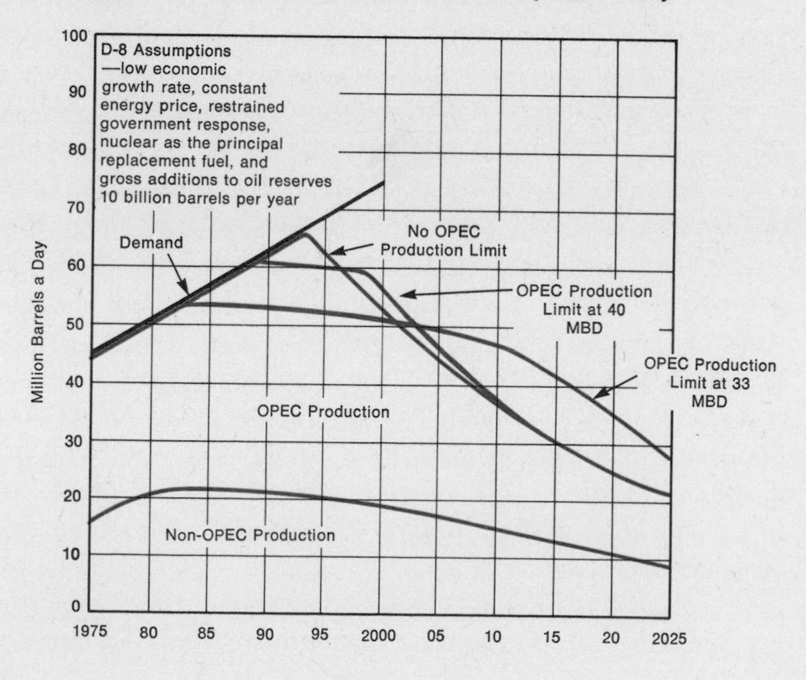

will be filled by imported oil. These forecasts require revision, and the implied national policies need reexamination. Figure I-7 shows that even assuming production is constrained only by a fairly liberal 15/1 ratio of remaining reserves to production level, production would peak at about 86 MBD.

But the rapid increase and decline of production in this profile makes it unrealistic. It seems unlikely that such a peak and decline would be justifiable on either technical or economic grounds.

The OPEC countries may decide for economic, conservation or other reasons to limit their production to a level below their future projected theoretical maximum. They might, for instance, choose a level of 40 or 45 million barrels a day depending on the rate at which their reserves are increased.

Such levels could probably be maintained for a few years without running into technical limits. With no limits on production by countries outside OPEC this produces not a peak but a plateau (at around 70 MBD in Case C-1) that stretches from 1989 to well beyond 2000, before technical limits (reserve/production ratios) force the curve down. This may seem comforting but in some ways it is alarming: the adjustment period to this lower production level for consumers (and producers of alternative fuels) begins several years sooner.

Production by OPEC countries of 45 MBD would require Saudi Arabia, which owns more than a quarter of all known oil reserves, to be prepared to increase output from its present rate of about 9 MBD to as much as 20 MBD. There is uncertainty about the level to which Saudi Arabia will wish to increase its rate of oil production, particularly when the financial and investment implications of such levels are assessed.

The third curve in Figures I-7 and I-8 shows the levels of production for the World Outside Communist Areas which would result if Saudi Arabia and certain other countries of the Arabian Peninsula were to restrict production to only slightly above present levels. This could mean maximum production by OPEC countries of 33 million barrels per day. Such a policy decision, if taken, could lead to a leveling of world oil production in the early 1980's and thus to a failure to meet the projected demand much earlier than has been generally expected.

Such an early limit could impose severe strains on all consuming countries. It is almost certain that the resulting shortfall in energy supply would lead to a significant price rise. This might trigger a world recession that would increase unemployment in many or all industrialized countries and severely damage the economies of the developing countries. These problems would arise not so much from a high price level as such but rather from the suddenness and the steepness of the price increases. In itself, a high-cost energy world could be as prosperous and appropriate for economic growth as a low-cost energy economy; it is the rapid transition that leads to the problems.

Demand

Our findings about prospective energy gaps and the special problems of oil rest on other WAES analyses, including detailed studies of energy demand and the possibilities for energy savings.

WAES has made a detailed analysis of energy use in 1972, 1985 and 2000 for 13 of the WAES countries* and has made estimates for all other regions in the world outside Communist areas (WOCA). Each of the national energy studies reflects a particular set of national conditions, priorities, and policies consistent with the appropriate WAES assumptions. It is impossible to summarize here the large differences in national energy demand patterns—though there is high learning value in studying each national projection. Here, we can only indicate some of the overall results.

Our projections for WOCA show primary energy demand growing from 81 MBDOE in 1972 to between 160 and 207 MBDOE in 2000—a range of about 25%. The upper end of this range lies well below the level that would be reached if historical demand growth rates were to continue; the lower end of the range lies somewhat above a level that might represent, by the year 2000, a zero annual energy growth. We believe our estimates to bound a plausible range, given our sets of assumptions (see Figure I-2).

Our projections show that energy demand will continue growing with economic growth, although more slowly than in the past. The slowdown is due to several factors—including saturation effects

---

* See the chapters for Canada, Denmark, Finland, France, Federal Republic of Germany, Italy, Japan, Mexico, The Netherlands, Norway, Sweden, U.K., and U.S.A. in *Energy Demand Studies: Major Consuming Countries* (MIT Press, 1976) and in *Energy Supply-Demand Integrations to the Year 2000: Global and National Studies* (MIT Press, 1977).

in some uses especially evident in the more advanced economies, the incentive to use energy more efficiently (induced in part by higher oil prices), and vigorous government policies to hold down consumption by improvements in efficiency where price alone is not adequate. Nevertheless, our projections show only a relatively small decrease in the ratio of growth in energy use to growth in GWP (Gross World Product), even with the substantial efficiency improvements assumed in the scenarios. This ratio is estimated to be from 0.82 to 0.87 on average between 1972 and 2000, compared with the recent (1965-1972) value of 1.02. If we look only at the industrialized areas of Western Europe, North America, and Japan, however, we find a more significant decline (with ratios from .77 to .80) than for the WOCA average. The ratio is somewhat higher for the developing countries with ratios between 1.10 and 1.05, compared to a 1960-1972 value of 1.26. The relationship between energy growth and economic development is an important and complex topic that deserves greater in-depth study.

Our demand estimates result in increasing energy consumption per capita in even the low-economic growth cases.

Developing countries' energy demands are projected to increase as a share of WOCA primary energy demand, from 15% in 1972 to as much as 25% by 2000. This estimate is based on a special World Bank Study made available to WAES.

Our demand projections are initially of *delivered* energy—energy delivered to the final consumer in such usable form as heating oil, gasoline or electricity. This is not the same as *primary* energy—the energy content of fuels before they are processed and converted. (Primary fuels are crude oil, coal at the mine, natural gas at the wellhead, and so on.) Our projections of delivered energy include the efficiencies of consumer devices. Energy processing losses are calculated in our supply-demand integrations, to produce the total primary energy demand figures cited above.

Figure I-9 shows the delivered energy demands by sector and the processing losses for two of the WAES cases. These losses grow from 26 percent of primary energy demand in 1972 to over 30 percent in 2000. This is due largely to projected increases in electricity use—growing at some 4 to 5 percent per year (compared to historic rates of 7.5 percent per year)—with an assumed generating efficiency of 35%. Expansion of district heating and cogeneration of

Figure I-9 Energy Uses by Sector, WOCA

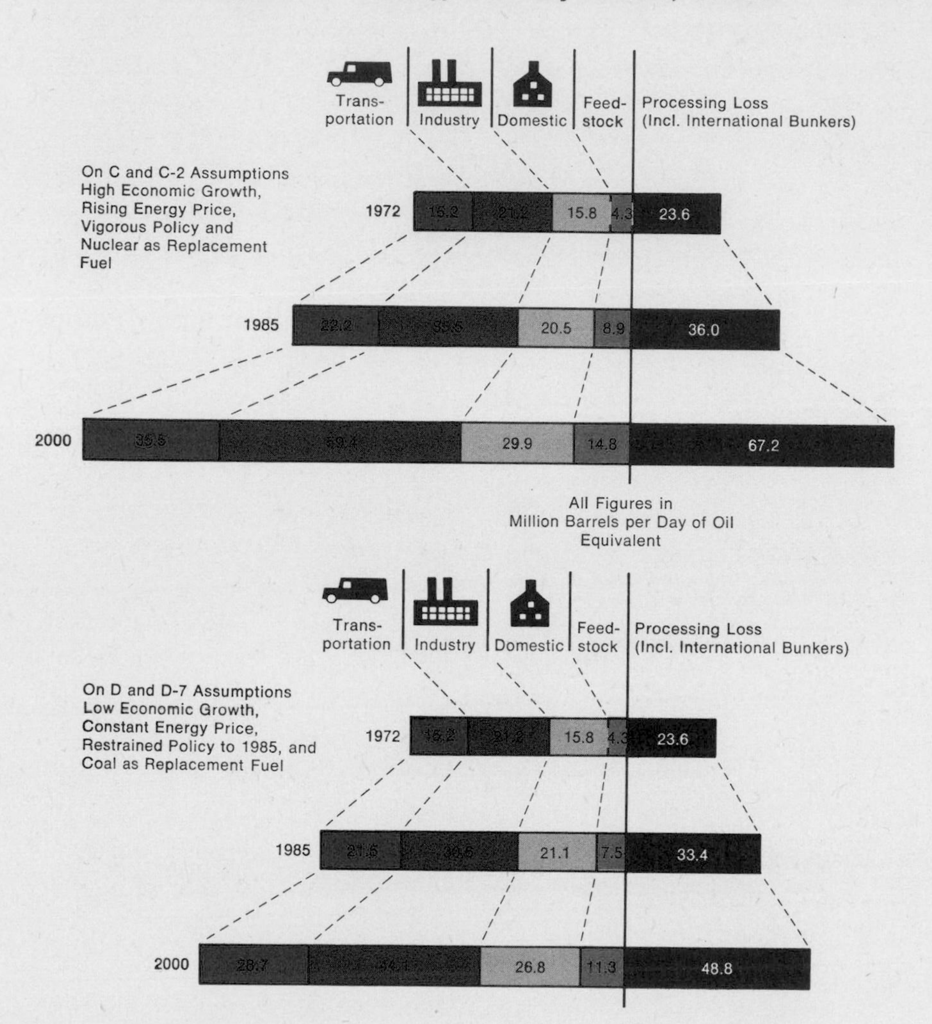

process steam for industry could improve this conversion efficiency by year 2000.

Figure I-9 also shows that our projections do not substantially alter the fractions of the total that the different activities represent. Yet some differences do appear. Transportation energy use declines from about 27% to about 25% of total energy demand by 2000, owing largely to automobile efficiency improvements and improvements in load factors in other forms of transportation. Industrial energy uses and fuels used as petrochemical feedstocks increase faster than other sectors and together account for over 50 percent of delivered energy demand in 2000. The share of residential and commercial demands do not change appreciably, reflecting (among many other things) the offsetting factors of reduced population growth rates in some regions and substantial increases in dwelling size and comfort in others.

Our demand projections include estimates of preferences for certain fuels. Figure I-10 illustrates the preferred demand for primary fuels in WOCA for two of our cases. Clearly, oil will continue to be the preferred fuel to the end of the century. In the year 2000 about 50% of oil consumption will go to transport and petrochemical feed-

Figure I-10 Primary Energy Demand by Fuel Type, WOCA

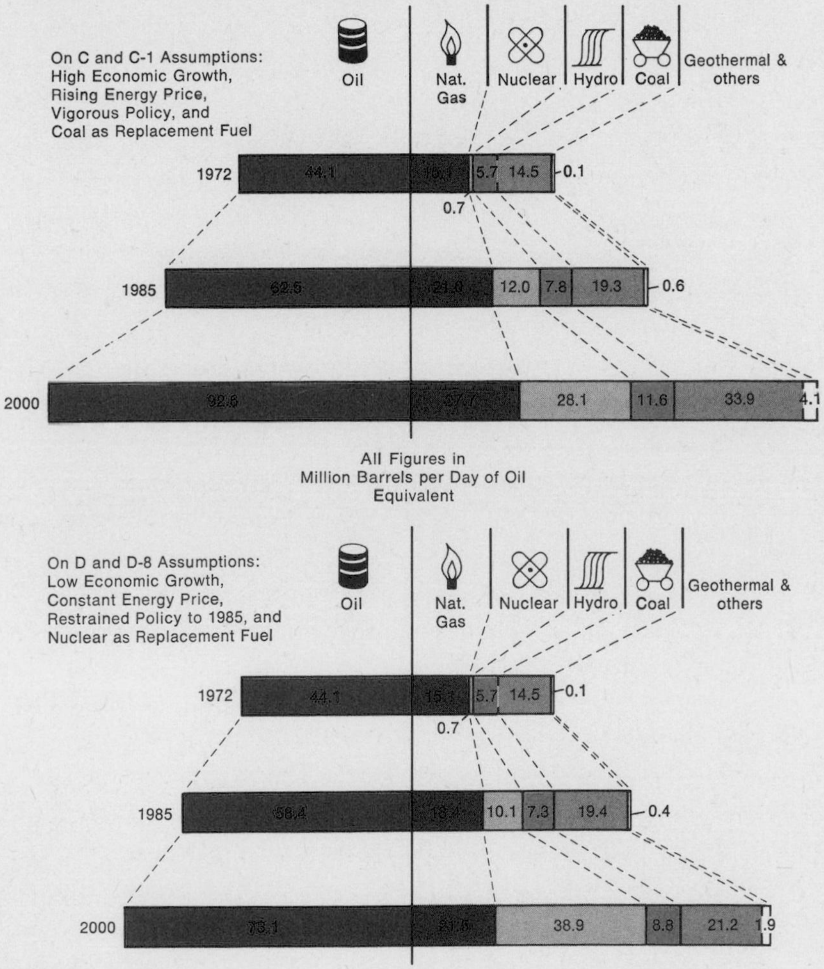

On C and C-1 Assumptions: High Economic Growth, Rising Energy Price, Vigorous Policy, and Coal as Replacement Fuel

	Oil	Nat. Gas	Nuclear	Hydro	Coal	Geothermal & others
1972	44.1	15.1	0.7	5.7	14.5	0.1
1985	62.5	21.0	12.0	7.8	19.3	0.6
2000	92.6	37.7	28.1	11.6	33.9	4.1

All Figures in Million Barrels per Day of Oil Equivalent

On D and D-8 Assumptions: Low Economic Growth, Constant Energy Price, Restrained Policy to 1985, and Nuclear as Replacement Fuel

	Oil	Nat. Gas	Nuclear	Hydro	Coal	Geothermal & others
1972	44.1	15.1	0.7	5.7	14.5	0.1
1985	58.4	18.4	10.1	7.3	19.4	0.4
2000	73.1	31.5	38.9	8.8	21.2	1.9

stocks, where it is an essential fuel. Even in these sectors, it will be necessary to use it in the most efficient way possible. In other sectors, substitution by other fuels is possible and will progressively take place.

Conservation

Our studies lead us to believe that much energy can be saved. Indeed, energy conservation may well be the very best of the alternative energy choices available. Its advantages and benefits are substantial.

But conservation is complex because both the opportunities and constraints vary widely among countries and among energy-using sectors. In some cases energy savings through improved efficiency follow from increases in energy prices; in other cases they require government action. We offer here some general examples drawn from our specific studies of the scope for holding down fuel consumption. Further discussion can be found in Chapter 2, "Demand and Conservation." Details about assumptions in WAES national studies on demand and conservation can be found in *Energy Demand Studies: Major Consuming Countries* (MIT Press, 1976) and *Energy Supply-Demand Integrations to the Year 2000* (MIT Press, 1977).

Most of the estimates for the transportation sector show significant improvements in energy efficiencies. For example, our projections for the average fuel efficiency of automobiles in the U.S.A. are 27 to 29 miles per gallon by 2000, compared to 13.5 mpg in 1972. Also, improvements in air travel load factors are expected in most countries, although the assumptions for increased air travel demand make this the fastest growing energy item in the transportation sector.

In the industrial sector, many national teams project improvements of about 1% per year in the ratio of energy use to value added in production to the year 2000.

The scope for energy conservation in buildings varies greatly from country to country, owing to differing needs, climates, and present standards. Yet substantial sectoral demand reduction from improved insulation standards and improved control and combustion equipment efficiencies—up to 30 and 40 percent or better in some cases—are projected, particularly in the U.S.A. and the Scandinavian studies.

These few examples are intended to illustrate the substantial amounts of conservation assumed in our scenarios. Energy conservation—measured as improvements in energy-use efficiencies—must be a starting point for rational energy policies. This is our main reason for isolating "national policy response" as a scenario variable. It is necessary, however, to distinguish between policies which reduce energy consumption by improving efficiency—maintaining social and individual requirements unchanged—and those policies which reduce energy consumption only by reducing economic activities. Measures taken by governments under pressure to reduce energy imports quickly should not be confused with the longer term objective of saving energy by improvements in efficiency.

Governments and industries have already taken important conservation steps, including fiscal measures, new regulations and standards, appointment of energy managers in companies, and so on. The considerable potential for conservation, although highly variable from country to country, depends upon the combined effects of energy price, government action, structural changes within the economy, and the product of innumerable decisions by many users, large and small. Thus, savings may not be so quickly realized as some believe or hope.

If substantial savings are to be made, continuing and intense commitments must be made and actions taken soon. While wide public approval for conservation measures may be relatively easy to obtain, many years are often required for such measures to reach full implementation because of their decentralized nature. Lead times are typically long. For example, it takes at least a decade to change over a stock of cars, 20 to 30 years for most industrial equipment, a century or more for a nation's entire stock of housing. Changes in a country's energy-using system are inevitably slow; they happen neither effortlessly nor overnight.

But energy conservation is essential. Decisions and actions are required by governments, industries and individual consumers if energy-saving measures are to be implemented. Conservation unquestionably must play a central role in global and national energy strategies to the end of the 20th century and beyond.

Coal

As oil production peaks, levels, and declines, substitute fuels will be needed on an enormous scale. Coal could be one of the major energy gap-fillers for many countries. In theory, the size of the resource base makes it possible to increase coal output significantly. But will countries want to produce coal and will others want to consume it on the scale required and within the WAES time frame?

Compared to oil, gas and electricity, coal has been dirty, awkward stuff to distribute and use. In developed countries its share in total energy has declined steadily as industries and private consumers have switched to more convenient and cleaner fuels, leaving coal in many countries with little more than the power station and steel industry market. Can this trend be reversed? There are certainly many technical possibilities. Coal can be substituted for oil and gas in electricity generation and for process heating in industry; it can be converted directly to oil or gas, though at substantial cost in both financial terms and energy conversion losses. In some countries coal has been turned into clean, smokeless and relatively easy-to-handle solid fuel for heating homes and offices. New methods of clean and smokeless combustion of raw coal currently under development need to be perfected. Improved techniques for the disposal of residual ash are needed. But the costs and lead times for building the conversion plants and changing user equipment and fuel preferences will be substantial.

Coal is abundant. Economically recoverable world proved reserves of coal are about 700 billion metric tons (equivalent to 600 billion tons of hard coal) or 3000 billion barrels of oil. Potential reserves are much greater, possibly as much as 12,000 billion barrels of oil equivalent.

However, known coal reserves are very unevenly distributed. Three countries (the U.S.A., U.S.S.R. and China) account for nearly 60% of present world coal production, with Poland, West Germany and the United Kingdom adding another 15%. There is also significant potential for coal development in the Southern Hemisphere. Historically, coal exploration has tended to cease when reserves sufficient to support local requirements have been found. In the future, reduced oil availability may provide an incentive for increased coal exploration and development. This could be particularly significant for developing countries which can exploit indigenous coal resources to reduce their need for oil imports. Additionally, potential coal exports could be a source of foreign exchange earnings for some countries.

Thus, abundant reserves of coal are probably available in many countries around the world. The pertinent question is whether coal will be produced soon enough, given the long lead times, the large financial investments involved, the need to attract more manpower for deep mines by improving working conditions, and the increased productivity resulting from technological developments. Our analyses show that with the given assumptions about energy prices, economic growth and national policies, coal production in WOCA could expand threefold to 3.2 billion metric tons a year by 2000 with 1.8 billion tons in North America. Figure I-11, which shows all fossil fuel production for one high-growth case, illustrates the potential dramatic expansion of coal compared to oil and gas.

As with all our supply projections, however, these figures reflect only potential production. There is no way of predicting how much will actually be produced, for that depends on whether, when,

Figure I-11 Potential Fossil Fuel Supply

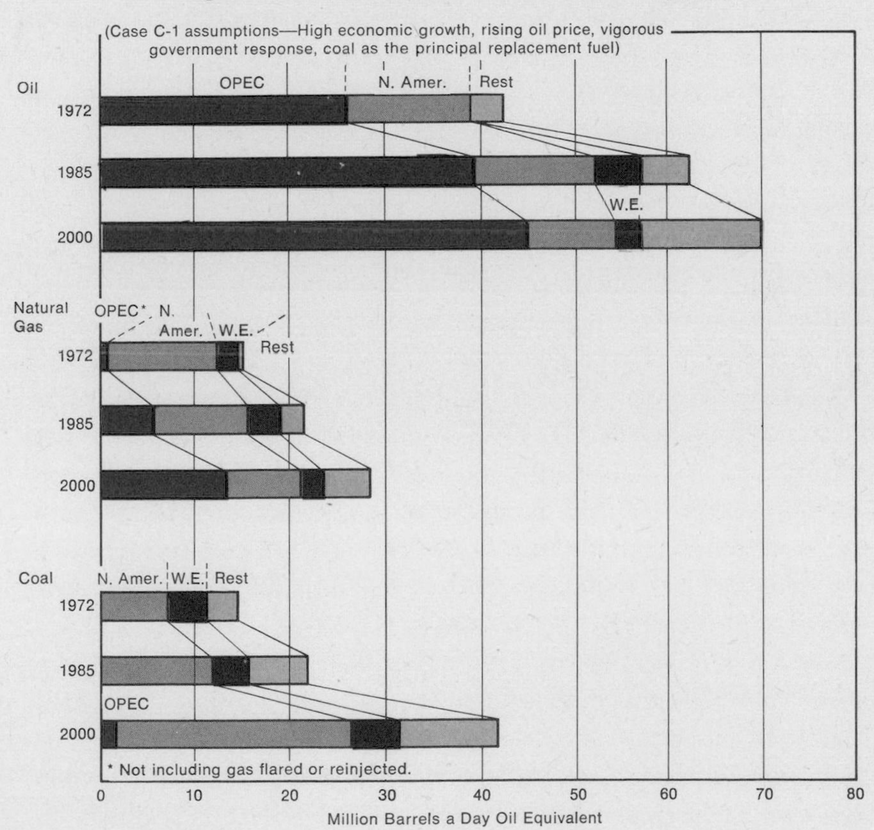

and on what scale decisions and commitments are made. To reach such high production levels, major and early investments will be needed for mines, coal-handling equipment, transportation systems, and devices to use the coal. Our detailed study of U.S. coal production potential, outlined in Chapter 5, shows that by year 2000 significant new and expanded transportation facilities (coal unit trains, convential trains, rail lines, coal barges, coal trucks, and coal slurry pipelines) would have to be built to get the coal from the mines to the markets and to ports for possible export.

Environmental standards must be maintained if coal is to be a major alternative to oil. A better understanding of the long-term climatic effects of fossil fuel combustion is essential. Surface-mined areas must be reclaimed and sufficient water made available for mining operation and land restoration. In several WAES countries local governments are concerned about the added costs of social services for large influxes of new workers in sparsely populated regions. Such environmental, economic and social issues require resolution before significant mine expansion, increased domestic use and large coal exports would be acceptable.

Our studies show that with coal, as with oil, there is a major conflict between desire and reality; but the conflict produces opposite effects. National preference for oil produces a prospective shortage. National distaste for coal produces a potential surplus. Our projections of desired coal demand show that total coal use will be well below the level that energy-consuming countries would have to reach to avoid an overall energy supply gap. This problem has already been illustrated in Figure I-3.

Large increases in coal demand require action to encourage its use and prevent excessive air pollution. Distribution systems must be built, new coal-fired equipment designed and installed, and techniques for controlling air pollution substantially improved. Coal's price in the market after providing for the cost of new transportation, coal handling, and pollution control facilities—must be sufficiently below that of competing fuels to encourage its use.

A major long-term potential for coal lies in its conversion to oil and gas. This option avoids the conversion of existing oil and gas distribution and consuming systems and may simplify pollution control by concentrating coal use at a small number of sites. There are large technical problems which can probably be solved in the medium-to long-term through research, development and demonstration. Technological improvements such as underground gasification of coal and economies of scale can probably decrease the present high cost of coal liquefaction and gasification. But synthetic fuels often require government action since very few companies are willing to undertake the financial risks without appropriate government guarantees at this stage of development. The time available between now and the end of the century may be insufficient for any major impact of coal conversion technology on the supply-demand picture by the year 2000. Chapter 5 provides a more detailed discussion of the potential production of synthetic fuels from coal and possible constraints on their development.

The scale of the coal challenge is as immense as the potential benefits. Above all, it is a challenge to our ability to anticipate and collaborate. Coal will not be produced in time, nor in sufficient quantities, unless firm commitments are made soon in potential exporting countries. It will not be exported in sufficient quantities, in time, unless energy-consuming countries see the need for coal imports well in advance and place firm long-term contracts. But this may not occur unless potential users believe that the coal will be available and that it can be burned cleanly.

Natural Gas

Natural gas is a clean and convenient fuel, well suited for household and commercial uses and for special industrial applications which exploit particular physical and chemical properties of gas. World reserves are large and unlikely to limit production within the next 25 years. However, the future role of natural gas as an energy source will be determined, not by the resource base, but by the problems of transporting and distributing gas from the wellhead to the consumer and by the attitudes of producers toward export.

Natural gas has historically moved directly from supplier to consumer through pipelines. The expense of constructing costly pipeline networks could only be justified where there were both large reserves and assured demand. Dependence on natural gas as a percentage of total energy varies from virtually zero in Sweden, Denmark and Japan, to around 30% in the U.S.A. and to 47% in The Netherlands where reserves are large.

An alternative to pipelines is to transport liquefied natural gas (LNG) in special refrigerated tankers at −161°C and regasify it at receiving terminals. This requires large capital investments at every stage, and about 25% of the gas is lost in the processing and transport of LNG. Capital cost estimates for such a system to carry gas from the Middle East to Japan or to the U.S.A. range upwards from $10,000 per daily barrel of oil equivalent. Because it is a clean and convenient fuel this cost is competitive with other fuels in certain important applications.

Large numbers of LNG shipments pose a special environmental problem. While there are risks inherent in all energy supplies, there is special concern about the consequences of an LNG tanker collision in port. Although many measures are being taken to ensure safety and further research is being done, this concern could slow the growth of LNG shipments.

Conversion of natural gas to methanol simplifies the transportation problems by eliminating the need to build complex and expensive LNG tankers and related liquefaction and regasification facilities. However, production of methanol requires building large conversion plants in the gas-producing areas and involves a 40% energy loss in the process as compared to 25% for LNG. Comparative calculations based on present costs indicate that conversion of natural gas to methanol could be competitive with LNG for ocean transportation of distances greater than 10,000 kilometers.

The world's major gas reserves are sited far away from present and potential markets, especially markets in Western Europe, Japan and North America. Because of the transportation problem as well as political factors, intercontinental trade in natural gas has been slow to develop and in 1975 was only .3 MBDOE. Nevertheless, gas is such an excellent, low-pollution fuel that much effort is warranted to overcome problems which limit large growth.

Figure I-12 shows the striking contrast between the location of natural gas reserves and consumption in 1975. North America has the most conspicuous role, with a consumption of some 68% of the gas used in WOCA. Such large usage has been based on domestic reserves which are now being rapidly depleted. Large usage has been stimulated by price regulation as in the U.S.A. In Western Europe the increase in gas consumption has been very rapid in the last decade. The major Groningen and southern North Sea fields are likely to be

Figure I-12 Proven Reserves and Consumption of Natural Gas in 1975 (In World Including Communist Areas)

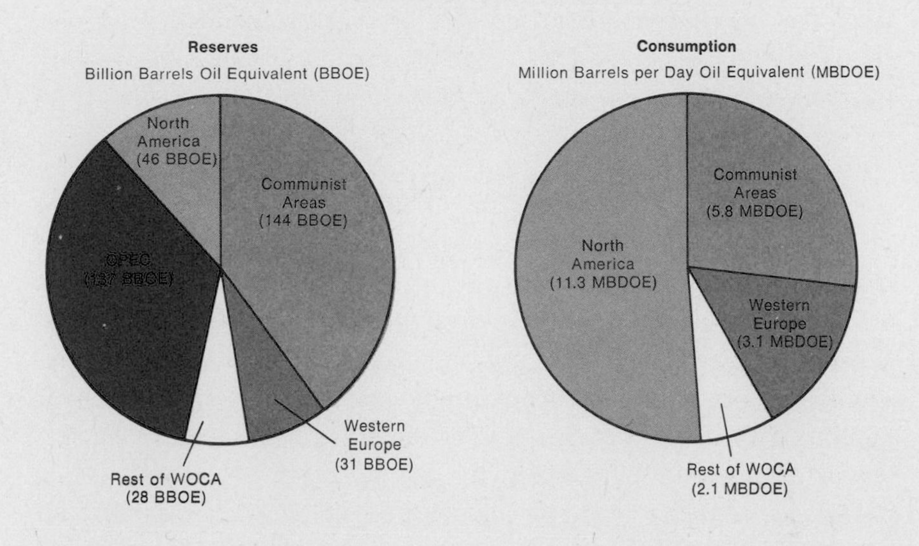

declining in the 1980's, although reserves in the northern North Sea could postpone the decline of overall Western European production at least to the 1990's. Japan, which lacks any significant reserves, is developing a natural gas industry based on imported LNG in order to diversify its energy sources and to take advantage of a relatively pollution-free fuel.

Our estimates indicate that by the year 2000 natural gas import requirements in the high-growth case could be 3 MBDOE for North America, 4 MBDOE for Western Europe and 1.5 MBDOE for Japan—or an import total of about 8.5 MBDOE.

To meet these import demands major commitments must be made to increase the pipeline and LNG systems beyond the maximum planned imports of 4 MBDOE for 1985. Exports from U.S.S.R. into Western Europe could contribute 1.0 MBDOE by 2000 but further increases are unlikely due to rising domestic demand for natural gas from the U.S.S.R.'s own reserves. The remaining import requirements of about 7.5 MBDOE would have to come largely from OPEC countries. Figure I-11 shows the large OPEC production needed to meet growing domestic uses and to satisfy desired gas imports by major consumer countries.

From a resource perspective, potential production by OPEC

countries could meet the needs of the major consuming countries in year 2000. Uncertainties about the growth of world gas trade stem from the attitudes of the governments of OPEC countries toward export of gas versus domestic use, including use as a chemical feedstock, the availability of capital for investment in LNG systems, and possible repercussions of an LNG tanker accident.

Nuclear Energy

Some people believe that nuclear power, by supplying a large fraction of the world's electricity, will substantially ease the pressure on fossil fuel resources by the end of the century. Their confidence is based on the relatively low cost of nuclear electricity and the outstanding safety record of operation of nuclear reactors and related facilities over the last 25 years. Yet the unique character of nuclear energy as a source of radioactivity and its potential for destruction has led to resistance in many countries to the growth of the nuclear power industry. The public debate on such issues is widespread and goes on in many countries at different levels of intensity. There are several serious concerns around which debate has turned. Probably the most serious is that of containment of radioactivity. The safe transport, storage and treatment of spent fuel elements and the resulting highly radioactive wastes which remain active for hundreds and thousands of years are matters of great concern.

WAES did not attempt to seek agreement on how and when the nuclear debate would be resolved in various countries. Instead, we decided to show the scale of potential contributions to the world's primary energy needs in 1985 and year 2000 that could come from an intensive and continuing nuclear power program. In this way, we could show how the pace of nuclear development would affect the total supply of primary energy required to meet the world's future energy needs.

In order to project the maximum contribution from nuclear energy we asked each WAES national team to estimate, in the general context of our scenarios to 1985 and 2000, the *maximum likely* and *minimum likely* levels of installed nuclear electric capacity. The high nuclear level assumes that such energy will be the principal replacement fuel for fossil fuels and that present political and technical obstacles will be quickly overcome. The lower nuclear level also assumes early resolution of obstacles which now delay nuclear growth in many parts of the world, but it is based on coal as the principal replacement fuel for oil. In many cases these high and low projections are based on national forecasts. After adding the national estimates we estimated growth in the rest of the world to get our totals. In this way we estimated what nuclear energy might contribute to world primary energy in the year 2000. Such estimates also allowed us to show how much energy from other fuels would be required if the supply of nuclear energy fell short of such WAES projections.

Each reader can form his own conclusions about the likely outcomes and the probability that the WAES nuclear projections will be achieved. If nuclear energy supply turns out to be lower than the WAES projections, then it will be necessary to consider carefully how the shortfall might be met—from what other fuels, or with what additional conservation measures.

The *maximum likely* and *minimum likely* WAES projections of future installed nuclear capacity to year 2000 are shown in Table I-2. The *maximum likely* projection would require an annual growth rate of 14% per year for 25 years. The *minimum likely* projection would require an annual growth rate of 11%. While these numbers may seem high, it must be recalled that the potential growth is from a comparatively low base.

Such rapidly rising levels of nuclear energy would permit this source to supply as much as 21% of the world's primary energy in

Table I-2 Summary of WAES Estimates for Installed Nuclear Power Capacity

	1974	*1985*		*2000*	
	(Installed Capacity)	*Maximum Likely C*	*Minimum Likely D*	*Maximum Likely C-2*	*Minimum Likely D-7*
Nuclear Capacity GW(e)[1]	66.9	412	291	1772	913
Percent of Primary Energy in WOCA	2%	9%	6%	21%	14%
Oil Equivalent (MBDOE)[2]	1.7	10	7	43	22

[1] GW(e) = 1000 Megawatts (electric) = 1×10^9 watts (installed capacity) for a power plant.

[2] We have adopted the convention of expressing nuclear electricity production in terms of the fuel input that would be required to produce the equivalent amount of electricity output from fossil-fueled power stations. The assumed generating efficiencies are 35% in 2000. The formula used to convert GW(e) to MBDOE, assuming 35% efficiency and a 60% load factor is:

$$\text{GW(e)} \times \frac{8760}{620{,}000} \times \frac{1}{.35} \times .6 = \text{MBDOE primary energy input}$$

2000, an amount equivalent to 43 MBDOE, as much energy as is contained in total WOCA oil production in 1975 or in 3 billion tons of coal per year. Even with this sizeable nuclear contribution, over three times as much primary energy in 2000 would have to be met by other fuels.

Our analyses suggest that with political will and firm, prompt actions on the many technical problems, WAES projected nuclear expansion is technically feasible in terms of uranium, physical facilities, and other critical resources. To achieve such nuclear growth, reactors would have to be built and uranium exploration and mining would have to be rapidly expanded, as would uranium enrichment facilities and fuel reprocessing facilities. Further research and demonstration is required before the most acceptable reprocessing methods and the most suitable sites for waste disposal can be selected. If these things are not done then our projected nuclear expansion might be delayed several years.

The nuclear issue does not have to be an all-or-nothing choice. There are in fact three stages of choice in a nuclear program: (a) reactor operation which involves a single use of uranium without fuel reprocessing, (b) the processing of used fuel to extract and recycle plutonium and uranium, and (c) operation of fast breeder reactors. Much of the nuclear debate centers on parts (b) and (c); very little of the debate centers on part (a). Yet in some places approvals to go ahead with part (a) and build nuclear power plants are withheld because issues in parts (b) and (c) are unresolved. Separation of these stages of choice could clarify issues and might allow nuclear power plant construction to continue (with its 6-10 year lead time) while the issues of parts (b) and (c) are being resolved. Some countries have already selected a course of action; others will have to choose.

Beyond 2000, other challenges that must be faced in the next few years appear. Our high nuclear expansion can be maintained only if fast breeder reactors—which both "burn" and produce nuclear fuels—become commercial around 1990-2000. At the same time, very large additional reserves of uranium must be found and developed. France, the U.K., the U.S.A., Japan, and the Debenelux venture (Germany, Belgium, The Netherlands and Luxembourg), and an international venture comprising electricity companies from France, Italy and Germany are developing fast breeder reactors; and in the

U.K. and the U.S.A. progress is very slow and surrounded by controversy. The technology is new and the lead times long; the scale of expansion by 1990 is thus uncertain.

It seems unlikely that breeder reactors will provide more than 5% of nuclear energy in the World Outside Communist Areas by year 2000. Most people see little hope that nuclear fusion will provide anything before 2000.

Uncertainties surround all of our estimates of demand and supply to 2000. Because different countries may choose different nuclear policies, the range of uncertainty in our nuclear projection is greater than for other fuels. On the one hand, extended delays on nuclear programs in various countries could hold nuclear power to the levels projected for 1985, which are based on commitments and construction already under way in most cases. On the other hand, a new awareness of the imminence of a deeper and continuing energy shortfall arising from reduced oil supplies might lead to a public reappraisal of the risks and benefits of nuclear energy and a decision to accept the risks. All that we can do in this report is to show the scale of the contribution nuclear could make in 2000 and describe the issues in the public debate which will influence each country's political decision on nuclear risks.

Hydroelectricity

Hydroelectric power is a major energy source in many countries today, and it will continue to play an important role in providing electricity for tomorrow's world. Most of the future major hydroelectric expansion is likely to be in the developing areas of the world —where it is estimated that only about 4% of the potential sites have been developed. Although the prospects for hydroelectric expansion are vast in the developing countries, growth is somewhat constrained by the location of potential sites and by the long lead times involved with construction. Therefore we have made conservative estimates of the future growth of hydroelectric power in the developing countries, assuming primary electricity production* to increase from about

* We have adopted the convention of expressing hydroelectricity production in terms of the fuel input that would be required to produce the equivalent amount of electricity output from fossil-fueled power stations. The assumed generating efficiency is 35%.

1 MBDOE in 1972 to a maximum production of about 4.5 MBDOE in 2000. This estimate could be higher if a number of developed countries, because of energy resource shortages or environmental reasons, choose to transfer some energy-intensive industrial activities such as aluminum production to those developing countries which have abundant hydro resources, access to required raw materials and the wish to develop such industries.

Growth of hydroelectricity in the developed world—which presently accounts for over 80 percent of WOCA hydroelectric capacity—will be lower than in the developing countries because the most favorable hydroelectric potentials are already developed and because environmental concerns may limit development of some potential sites. New technologies such as axial-flow turbines might extend the number of potential sites. Our projections show a primary hydroelectric growth in Western Europe, Japan and North America from 5 MBDOE in 1972 to a maximum of 7.5 MBDOE in the year 2000.

Other Fossil Fuels

The other fossil fuels—heavy oil, oil sands and oil from shale—are of immediate interest because they can be converted into liquid fuels similar to those obtained from conventional crude oil and can be fed into the existing energy infrastructure. The resources of such fossil fuels are large compared with resources of conventional oil, but current production is less than 0.2 MBDOE. Our analyses show that in our rising energy price cases their production could be 3 MBDOE by the year 2000. The capital and operating expenses to produce such oil are greater than for conventional oil, and developing some of these sources presents significant environmental problems which must be resolved before production is greatly expanded. World oil prices may have to be higher than they are now to stimulate large increases in production of these fuels.

Heavy oil and oil sands are two major "other fossil fuel" reserves found principally in Venezuela and Canada. The resource base in Canada and Venezuela total about 2,000 billion barrels of oil, but only 700-800 billion barrels are considered to be ultimately recoverable. In Canada about 30 billion barrels of oil may be recoverable using surface mining methods. The remaining bulk of the vast resource will be available only through in situ recovery methods. Before there is large-scale production, serious social, financial, technical and environmental problems must be solved. Projections in the WAES scenarios indicate a possible Canadian production of 0.8 MBD by year 2000. The resource base implies that production could be many times these levels, but this would require very vigorous technical and financial action on a scale not yet in evidence.

The largest known oil shale reserves are found in the United States, with significant amounts also in Brazil, the U.S.S.R. and in China. Smaller quantities are also found elsewhere, for example in Sweden. Development activities to date suggest that various technologies could be used to extract oil from oil shale at a substantially higher cost than the current price of imported oil. Government subsidies for at least the early stages of development appear necessary, and environmental problems must be solved before large-scale production can occur. A maximum worldwide production of oil from shale of about 2 million barrels a day in year 2000, mainly in the U.S.A., is projected in our rising price scenarios.

Geothermal

Geothermal energy can be produced from natural steam resources or from hot dry rocks. Natural steam is economically competitive, but the resource base is limited since it requires the relatively rare geologic combination of hot rocks, an underground water system, and an impermeable caprock for trapping the steam and providing pressure. Total installed global geothermal electrical generating capacity was only 1,400 MW(e) in 1975. Energy from hot dry rocks could greatly increase geothermal resources, but the technology is still at an early stage of development.

Solar and Other Renewables

Solar heat, solar electric, wind, tidal and other renewable energy sources are beginning to attract research and development efforts because in the longer term they are clean, sustainable energy sources of considerable size. Certain applications, for example solar water heating and space heating, are now economically competitive in some countries and could become more so as other forms of energy become scarcer. For other solar technologies, investment costs are still unattractive, operating and maintenance costs uncertain, and there are difficult technical problems to resolve, particularly the storage of energy for sunless or windless periods. Further, there are marked differences between countries and even between regions only 100 miles or so apart in solar or wind "climates" and therefore in the cost-effectiveness of these sources.

These factors make it very difficult to assess how rapidly renewable sources will expand. The WAES national studies indicate a maximum contribution from solar energy (both thermal and electric) in the developed countries of about 2 MBDOE by the year 2000. This figure is uncertain; if the price of oil were to double in real terms, the contribution of renewable sources could be much higher. Balance of payments problems arising from imported oil costs may accelerate action. There is also the possibility of major technical and cost breakthroughs which could bring on these sources more rapidly than we have assumed.

Although the aggregate amounts of energy available from these sources may be relatively small in the near future, they can be very important in human terms because of the large number of people, particularly in the developing world, who could benefit from them. Other advantages may lie in the decentralization of energy production and use which such systems offer.

Research, development and demonstration of renewable energy systems should be given a high priority as soon as possible. They have a critically important role to play beyond 2000 as oil and natural gas decline further, particularly if nuclear and coal are limited by resource and/or environmental and safety constraints.

Energy Pricing

Many people believe that the price mechanism could play a major role in producing and allocating energy efficiently and effectively. Yet it is widely recognized that price is insufficient for a variety of reasons. WAES selected a range of world oil energy prices as one of the principal scenario variables. The results of our studies indicate the likelihood of some energy prices increasing in real terms.

It could be argued that a higher oil price than we assumed would close the gaps found in our scenarios. We have, however, not attempted to quantify this—largely because our rising oil price case also provides a considerable stimulus for energy conservation and fuel substitution which will involve hard and costly decisions. This problem is also related to other important issues such as balance of payments and inflation which we have not studied. Appropriate internal prices would reinforce this stimulus.

There is urgent national need for a careful examination of policies and controls which influence energy availability and consumption. Higher internal prices for energy could encourage a much more rapid development of fuel-saving techniques and alternate energy sources such as solar energy and oil sands. They could also make fuel substitutions, such as the costly conversion of coal to gas and liquids, seem more attractive.

Conservation and solar energy typically have large front-end costs, but low operating costs. Evaluated on present discount criteria, such energy projects are disadvantaged compared to most conventional supply and use techniques. The possibility of some form of subsidy could be examined to take account of full costs and benefits. There may be need for international agreement on this, since individual nations may jeopardize their competitive positions if they act alone. For the same reasons, tax incentives to encourage investment in conservation and solar energy should be considered.

Consideration might also be given to broad, early policy action on energy supply, use, and substitutions. One clear signal from our studies is that the adjustment process in the market, even combined with reasonably vigorous government policies, may not alone be sufficient to make the appropriate changes unless these policies take effect in a timely fashion. Policies on pricing, investment incentives, and changes in public investment would need to be implemented very soon because long lead times are required for the fuel substitution and energy conservation which are needed to extend the life of the oil era.

Environment and Climate

Throughout our work we recognized the important relationship between energy policies and environmental objectives as well as the need to consider particular energy alternatives and their costs in relation to projected environmental quality standards in each WAES country. Changes in such standards are different in each country so that generalizations were not possible. We therefore asked each WAES national team to build environmental quality protection into their scenarios for 1985 and year 2000.

Environmental deterioration from energy production and use falls broadly into two general categories: 1) local effects that can be controlled at a cost with suitable technology, and can be included in prices of the product or service and 2) regional or global effects that are very difficult or impossible to control.

Local effects include water pollution from mining, transporting fuels, and disposing of waste heat from power plants, and air pollution from auto emissions and from products of oil, gas, and coal combustion (such as particles, sulphur compounds, and oxides of nitrogen). Environmental degradation from land disturbance in mining and spills from tankers or offshore oil operations are largely controllable when the public is prepared to pay the cost, either in the price of the product or as taxes, and when national and international controls can be effective (as in the enforcement of regulations to minimize ocean pollution from tanker operation).

One global effect of energy use that is of some concern is the accumulation of carbon dioxide in the atmosphere as the result of fossil fuel combustion—the so-called greenhouse effect. On the other hand, the presence of particulate matter in the atmosphere might affect climate in the opposite direction. Too little is known about the complex interaction of these phenomena to say which is the true effect. Much more systematic research, both at the regional and global levels, is required.

Another topic of regional or global concern is the handling and storage of highly radioactive waste products from the nuclear fuel cycle. Consideration of methods to dispose of such products safely forms part of the current debate on nuclear energy.

Knowledge about nearly all of these possible regional and global impacts of energy on the environment is very limited. The problem is aggravated by the great difficulty of separating the effects of the environmental impacts of energy from natural causes and the difficulty of determining the exact nature of the chemical and physical processes from which future trends may be predicted. Clearly this is of great importance; some experts fear that the effects on the climate of burning fuel may become irreversible. In that case preventative action would be effective only if the problem is perceived and action undertaken early enough. If such fears are not groundless, our analyses show that the world energy problems will indeed become acute. There is no doubt that we need to know much more about these issues.

Uncertainties

We feel that our assumptions about a range of futures are plausible and the analyses based on them persuasive. Although they are not based on an extrapolation of the past, these analyses do assume that there are no big surprises. This may not be so.

Technological breakthroughs—more economical and energy-efficient processes for in situ production of oil from shale, oil sands, or heavy oil, an increase in the average recovery of oil from existing oil fields significantly beyond 40%, in situ coal gasification, and rapid development and distribution of low-cost solar technologies—could all influence our projections. But lead times are such that their major impact will come beyond the year 2000. We consider it wise to classify unanticipated bonuses of this kind as just that: bonuses. To do otherwise would be imprudent.

Nor can the possibility of discontinuities, political instability and disasters be ignored. We have not taken account of the possibility that the next twenty-five years *may* see runaway inflation in major consuming countries; prolonged, severe, worldwide depression; sudden, large energy price increases or interruptions in the international energy trade. We hope that such events do not occur, but clearly if they did they would have serious repercussions on our conclusions.

We have included in our scenarios for 1985 and 2000 many vigorous assumptions on future government action that would promote energy supply expansion, conservation, and fuel switching; and these may not happen. Policies and actions anticipated in these scenarios are not self-executing—hard and costly decisions must be made and actions begun soon if the increases in energy supply and reductions in energy demand projected in our scenarios are to be realized.

National and International Issues

Energy is but one of the many important topics in both the expanding international dialogue among nations and the domestic debate within nations. Our purpose in WAES has been to define and analyze what seem a plausible range of world energy futures to the year 2000 and to provide a framework for discussions of the choices available if a balanced economic development of the world is to be maintained. While we confine our discussion to specific energy issues, we recognize that they are but part of a much larger national and international agenda.

Our analyses make clear the increasing importance of energy over the next few decades. They show that there are many opportunities to avoid the consequences of potential energy shortages if people and nations face the problems openly and together—and do so now. What is at stake is world economic stability. Nations will have to decrease dependence on oil and shift to other energy sources in order to sustain economic and energy systems into the 21st century.

The transition will be affected by a myriad of individual decisions which will be made in the context of national and international policy. These decisions will consider specific issues and the varied balances of advantage and disadvantage that surround each alternative. How and when should this coal mine be opened? What incentives can be used to insulate more houses? What can be done, and how quickly, to increase the energy efficiency of vehicles, and what are the implications for air-pollution standards? Should a nation limit oil production? Should it export its oil, natural gas, coal and uranium? Thousands of such decisions must be made, each with difficult balances to be struck between costs and benefits, employment, inflation, balance of payments and a host of other factors.

Our analyses make clear the critical and continuing energy interdependence among nations. Fuel resources are very unevenly distributed; thus a majority of nations—including most of the developing countries and many very large users—must depend on the decisions of fuel-exporting countries for growth of their economies. Oil is now the principal world fuel and the OPEC countries have a dominating influence on world oil prices and production levels. Some oil-producing and exporting countries have linked agreements on oil prices and production levels to progress in the negotiations on the demands of developing countries for a "New International Economic Order"—negotiations designed to strengthen their relative position in trade, commodity prices, transfer of capital and technology and other issues.

Of all the importing countries, the energy policy followed by the United States is of decisive importance for the rest of the world, because of the sheer size of its requirements for energy compared with all other countries, the extent of its domestic resources, and its ability to pay high prices for energy.

In the future other national groups may become the center of critical decisions that affect fuel supplies for most of the world: for example, coal from North America or some Southern hemisphere countries; uranium from Australia, Canada and the U.S.; gas from the OPEC countries; and possibly even large-scale solar production of fuels such as hydrogen and methanol in less-developed countries with abundant land and plentiful sunshine. The U.S.S.R. and China may in the future decide to expand their export of fuels to some countries in the non-Communist world, if their resources permit it and it accords with their perceived interests.

In the short run there seem to be obvious and legitimate differences of interest between the long-term national objectives of the exporting countries and those of the importing nations. The developing countries, especially those at the early stage of industry, also need substantial energy growth to fuel their economic expectations. On the other hand, deep common interests between various groups of countries should override these differences, since all depend on a growing and prosperous world economy without rampant inflation, and since all countries are consumers of energy. A greater understanding of these complex interrelationships is perhaps the major challenge of the future. This could open a way to resolve differences to the benefit of all parties; a dialogue like the one which has been taking place within the framework of the Conference on International Energy Cooperation might be one way. Although the differences which make agreement on these issues so difficult are formidable, the incentives for all to resolve these differences are compelling.

To underpin these developments, new forms of international cooperation with some capacity and authority to act on behalf of nations may have to be created. Entirely new financial arrangements may be needed to protect commercial investments in international exploration, development and production projects. On the other hand, better means may be needed to protect the purchasing power (in real terms) for exports of OPEC countries against imported inflation and currency devaluation as well as to protect the value of surplus earnings invested in the industrialized countries. International arrangements may also be needed to assist developing countries in stabilizing their export earnings and paying for imported energy vital to their hopes for economic development.

Possibilities should be examined for much greater international collaboration in research, development and demonstration for difficult and expensive new energy technologies. Energy research and development is now going on in many countries and several joint research programs exist. These efforts usually stop short of the crucial demonstration stage needed to launch a new technology toward rapid commercial expansion. International collaboration must not only move a technology forward but also overcome economic and industrial and social barriers preventing its rapid spread. In many cases large-scale production should sharply reduce costs and create markets, but private industry is inhibited from investing because present costs are too high and the markets too small. A few large-scale demonstration projects backed by collaborative funding could break this stalemate and benefit many. Such projects could be carried out by national groups within a cooperative framework for planning, evaluation, and funding.

Conclusion

The Workshop has attempted to provide a coherent description of the world's energy system based on estimates for each country, and has suggested some of the measures that need to be taken. This report provides a framework for the continuing energy debate. No doubt its conclusions will be modified. This type of study should be pursued on a continuing international basis, so that individual nations may have access to a common understanding of the problems ahead and of the international consequences of national actions. However, in the immediate future the next steps have to be taken at a national level and it will be for the individual members of the Workshop, and others who read its report, to urge the necessary action on their own public and governments.

PART II

FRAMEWORK FOR THE ENERGY DIALOGUE

CHAPTER 1

MAPPING THE FUTURE: THE WAES APPROACH

The Challenges of Forecasting — The Overall Framework — The Scenario Variables — Selecting Useful Scenarios: Demand; Supply; and Supply-Demand Integration

The Challenges of Forecasting

Three thousand years ago Cassandra told the people of Troy that "the city will fall." No one listened because that was all she said. Prophecy is an idle art unless it stirs us to action. It has to be instructive in order to be useful; and it has to be credible.

Because the Workshop's principal purpose was to lay the groundwork for local, national, and supranational action, one of the main criteria for choosing methods was credibility. Just how, at a time of great uncertainty and rapid change, with an overarching issue as complex as energy, and in a world not of one city but over 140 interacting nations—each with different resources, constraints, opportunities and aspirations—can one hope to make credible forecasts on which to propose credible actions?

Furthermore, how can one do this in a sufficiently global yet adequately concrete way? Actions are usually about specific, local matters. Even national strategies must be fairly narrowly defined: not "we must produce more fuel" but "we must drill so many oil wells, build so many power plants, open so many coal mines—here, at this date, with these technologies." Yet no nation is an island, least of all on the energy scene. The actions of others can make or break national plans and forecasts—as the world has known too well since October 1973.

These were the kinds of questions we faced when WAES began to work three years ago. In order to look 25 years ahead realistically, we had to understand thoroughly—nation by nation—what happens now. On this base we then had to consider alternative global energy futures that all nations would encounter, yet also ask how each nation would behave in its unique fashion. And we had to consider the totality—how would these behaviors in turn affect the alternative energy futures?

To handle these problems our methods had to be mutually consistent so that all national teams could work independently but in a coordinated way. Only then could we have confidence in the totality of our results and the actions that could be motivated by them.

We also wanted our projections to be more than dead exhibits in the museum of futurology. They had to be dynamic, flexible, teaching aids from which we, and others, could learn and go on learning as times and situations change. Besides, we could not encompass every component of the vastly complex and interacting, global and national, energy, economic and environmental "systems"—let alone allow for all future eventualities. Others will want to check our estimates, change our assumptions, see what factors we may have ignored, add their own pieces of the puzzle. We hope they will. We have attempted to make our method and approach as explicit and reproducible as possible.

In short, we had to invent a forecasting framework that could meet all these demands. This chapter describes that framework and how we used it to develop our energy projections. We begin with a broad outline of the framework—including a seven-step description of our process—and then describe the assumptions used in the projections. These serve to state the essentials of our procedure. The final three sections of this chapter—the "Analytical Methods" sections—amplify, in considerable detail, the most important elements in our basic methodology.

In the end it is not, of course, the procedure we followed that matters most: it is what our methods tell us. Our main messages are summarized in Part I; they are elaborated in Chapters 2 through 8.

The reader interested in results, more than methods, can skip much of this chapter (particularly the final, "Analytical Methods" sections) and go directly to later chapters.

Yet we think that our methods are important, and that our message is substantiated, and supported, by our carefully defined methodology. We thus devote this chapter to a presentation of our approach—the why and how of our mapping of the future.

The Overall Framework

The real world of energy decisions is a highly interactive one, and too complicated to mimic in an exact forecasting model that would also meet the basic WAES objectives described above. One cannot make any such model consistent and explicit—or even calculate from it. So we had to construct an approximate model—a simplified conceptual framework—of the real energy world, shown in Figure 1-1.

Its essential feature is that it "decouples," or separates, those global factors not under control of individual, energy-consuming nations from those factors under such national control.

This feature is also central to the *process* used by WAES in its studies. Broadly speaking, we begin by defining general plausible future states of the World Outside Communist Areas*; we see how these will affect each region and nation; and then we add up these national consequences to see how they affect the global environment. If the result is an absurdity—say a huge shortfall of oil—then clearly that future is not a credible one. We may discard it. Or we may present it to show what follows from trends and assumptions that are now widely held to be reasonable. Or, we may try to adjust the assumptions to find reasons for and solutions to the absurd situations. It is from this adjustment that one learns what must be done, what strategies are needed, and how we measure the "vigor" and timing of the strategy decisions that are required.

---

* Members of the Workshop are energy-consuming nations from outside Communist areas. Although the U.S.S.R. and China are major world energy producers and consumers, their trade in fuels with the countries in the World Outside Communist Areas (WOCA) has to date been relatively small. In the WAES projections, we assume that this situation would continue to the end of this century.

Figure 1-1: The WAES Conceptual Framework for World Energy Futures

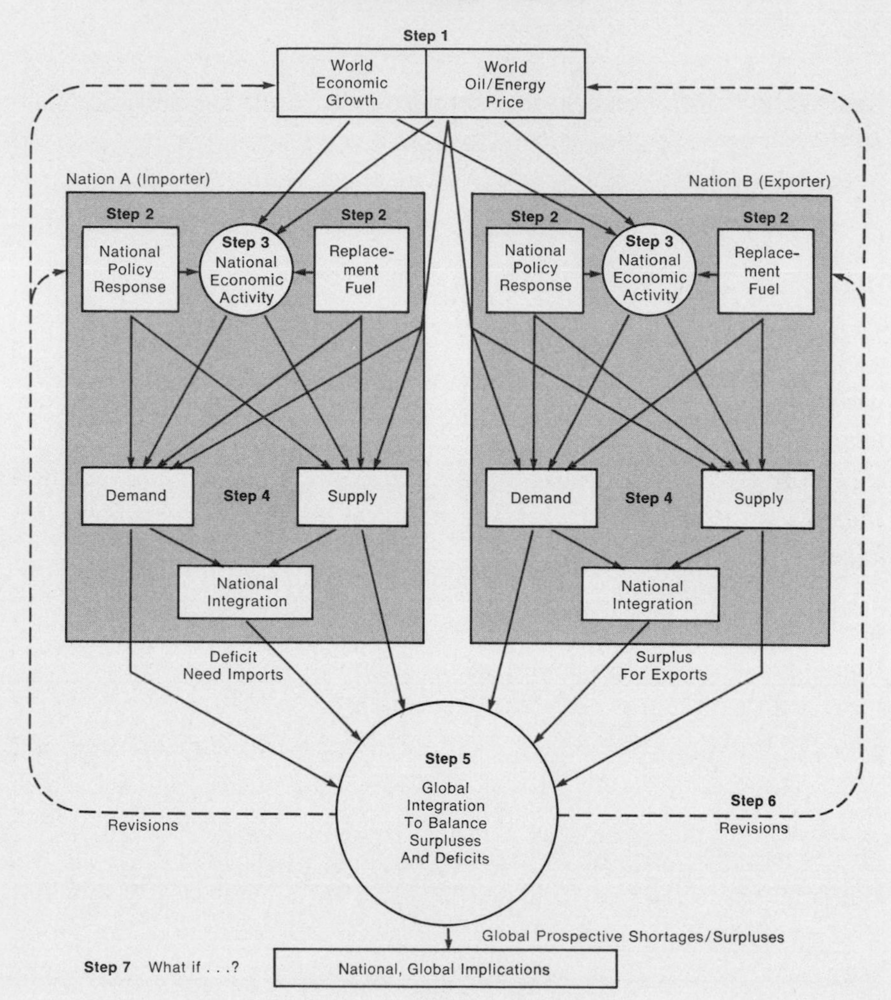

Our method involves seven basic steps. Roughly speaking, these trace through Figure 1-1 from top to bottom the following:

In our *first step*, we define in broad terms what sorts of "energy worlds" might evolve at our target dates of 1985 and 2000 by making varied assumptions about the key determinants of energy supply and demand that are independent of national decision-making processes: *world economic growth* and *the world price of oil or energy.* These are our two *global scenario assumptions.* They are not predictions, but plausible values that adequately span our assessment of the likely future states of the world—given the assumption that the past trends in world economic growth may be extrapolated into the future.

Scenarios, Predictions and the Real World of Energy

We use in WAES what has commonly been called the "scenario approach." That is, we do not predict the future; we are not prophets. Nor do we decide what the future "should" be like, and then try to get there. Instead, we attempt to identify a range of different but plausible futures. A "scenario" is a plausible energy future.

Scenarios are not forecasts; they are chosen to span a range of possible futures and lead to different estimates of energy demand and supply. We define and test several scenarios. For each scenario, values of key variables are specified and the consequent energy picture evaluated. When significant findings remain unchanged under several different scenarios, confidence in those findings is increased.

But the real world of energy decisions is not quite an easily defined scenario. In the short term thousands of individual decisions are made by political bodies, industrial companies, individuals, and so on. In the longer term countries try to adapt their strategies to present realities and what they expect of the future. All these activities are affected by changing technologies, prices and costs, size of known fuel reserves, economic activities, balance of payments considerations, government policies, and many other factors.

Most of these things are under national control. But nations are also affected in ways largely beyond their control by what happens elsewhere. "Gaming" goes on between countries, each seeking political, economic and social advantages. Who will "win" is never certain. Producer cartels and embargoes, fuel export quotas and tariff barriers, economic activity abroad, and a host of other factors in the global environment can also sharply influence each nation and prevent it from controlling its own destiny. The real world of energy is, in short, a highly complex and interactive one. Our scenarios can only roughly approximate this real world.

Step 2 introduces, as a third global assumption, factors that nations can control. One is the level of *public policy response* to the emerging energy situation. Individual nations must and will respond to changing world energy situations, and the responses may be of varying types. We assume that until 1985 each nation has either a *vigorous* or a *restrained* national policy response (for example, by enacting policies to bring on new supplies or to promote energy conservation). From 1985 to 2000 all energy policies are assumed to be vigorous. (These admittedly subjective terms are clarified later on in this chapter.) Also during the 1985 to 2000 period, a major question facing nations could well be framed as: What is the main replacement source for declining oil availability? We selected, as a fourth global scenario variable for this period, the choice of *principal replacement fuel*: coal or nuclear.

Step 3 moves from these global assumptions to corresponding regional and national scenarios. Each nation will move at different speeds, in different directions, even within the same global context. Economic growth rates will vary. So will energy demands and supplies and a host of related factors. So each national team made its own translation and application of the global assumptions to their own national economic assumptions (taking account of trade capability, industrial infrastructure and many other factors). These national estimates underwent continual modifications through criticism, comparison and discussion at frequent WAES meetings.

Thus we move from world and regional economic growth and energy price to national policies and activities. *Step 4* is the detailed analysis of national energy supply and demand within the context of the world, regional and national scenario assumptions. The calculations of supply and demand are done separately, producing estimates of desired demand and potential supply. *Desired demand* is what each nation would "like" to use (assuming supplies for each kind of fuel are available) at the given price, economic growth, and policy assumptions. Maximum *potential supply* is what could be produced within a country, given the scenario assumptions. These demand and supply estimates do *not* take into account the availability or unavailability of imports, or markets for potential exports.

National energy forecasts very often stop at Step 4 and simply assume that any shortfall between home-produced fuels and demand can be filled by imports. Who asks how all these imports add up—

or whether all that oil will be available from producer countries? If limited supplies must be shared, how much will each country get? WAES faced this question in *Step 5* of our approach. This step compares the fuel imports and exports of the national supply-demand integrations with global availability by means of an international balancing calculation.

We shall describe this *global integration* technique later in this chapter. For now it is only necessary to realize that we use two kinds of integration—*unconstrained* and *constrained.*

Unconstrained integration compares desired energy demand and fuel mix with projected potential supply to see what the shortages and surpluses might be. This integration is unconstrained, or unaffected, by the possibility of insufficient imports to meet preferred fuel demands.

In reality, of course, there can never be actual shortfalls of fuels over extended periods of time. Prices will rise, inducing a higher supply level and lower demand level, and/or policy actions will reduce demand to match supply. Yet consideration of these imaginary shortfalls, or "prospective shortages"—as well as "prospective surpluses"—can be extremely instructive. It indicates where problems are likely to be, and their possible size. It shows where to revise our expectations or focus our strategies.

Our second type of global integration, the constrained integration, forces substitution of available fuels for shortage fuels on a regional basis (when necessary and within the limits considered feasible). The aim is to reduce the projected shortages and surpluses by using all available fuels in the most efficient way possible, at the least total cost to consumers. This integration process is a reasonable approximation of the real world insofar as energy decisions over the long term, and within physical and political constraints, will be made on a least-cost, most-efficient basis. The integration is constrained by a series of limits on supply availabilities, which may not be sufficient to meet preferred fuel demands, and by specific ranges of demand preferences for major fuels.

From the global balancing integrations, we obtain an overall energy balance (or imbalance)—the total energy supplies and demands, shortfalls and surpluses, for each major fuel and for each scenario case, including the unconstrained and constrained alternatives. We can then evaluate each scenario for each country in terms of feasibility, consistency, and likelihood. For example, is the economic growth rate too high at a particular energy price for economic feasibility? Is there a prospective supply-demand gap? Are the environmental impacts of our assumptions too severe? And so on.

Step 6 in the WAES framework allows a major choice. We can either end our analysis here and present the results as a final scenario or projection. Or, if the projection points to very severe prospective energy shortages or other major problems—in fact to an "impossible" or highly undesirable future—we can try to revise that future by altering our assumptions or by calculating new scenarios. This revision process is shown by the dashed lines on Figure 1-1. In fact, as described later in this report, we used a mixture of these methods.

Step 7 lies outside our main conceptual framework. We ask, in effect, what would happen if our main assumptions proved incorrect. What if Saudi Arabia oil production is held close to present levels? What if solar energy rapidly becomes cheaper than expected? What if the oft-conjectured serious climatic impacts from burning fossil fuels are substantiated? And so on. Such questions are partly explorations of the *need* for strategies, but, also, they test the impact of major uncertainties on our scenarios. We discuss these issues later (in Chapter 8).

A large uncertainty in our work has been the possible future role of the U.S.S.R. and China as energy exporters or importers. Although the U.S.S.R. and China are major world energy producers and consumers, their trade with the rest of the world has to date been relatively small. We assume that their future role in world energy trade will continue to be small. Estimates by others, which we have adopted, indicate that major energy supply expansions will be needed in the U.S.S.R. and China to meet their growing domestic demand for oil, gas, and coal—thus leaving only marginal amounts for export. This is an uncertainty in our projections that other groups might like to clarify, and use the results to adjust our forecasts and conclusions.

We believe that with the main assumptions employed our scenarios are robust: slight changes in the key assumptions make little difference in the outcomes or scale and timing of necessary strategy responses. But major uncertainties—departures from the trend—are possible.

National teams from 13 of the countries within the Workshop

did the analyses—nations which together accounted for 78% of WOCA's energy consumption and 77% of global economic activity in 1975. We relied on a World Bank study for estimates of energy and economic growth in the developing countries and made our own estimates for non-WAES industrialized countries. Supply estimates for non-WAES regions such as OPEC, on which WAES nations depend for most of their energy supplies, were produced by special WAES studies for each fuel (oil, gas, coal and uranium) as described in Chapters 3, 4, 5, and 6. All of these analyses are described later in this report and in one of the WAES Technical Reports, *Energy Supply to the Year 2000: Global and National Studies.**

The Scenario Variables

Having defined the overall framework, our next task was to settle on the main scenario variables of energy price and economic growth. We had to choose quantities for 1985 and 2000 that gave a reasonable span of plausible futures. We also had to limit their number to make the work load manageable.

For both demand and supply projections, we have settled on the following global variables:

1. World Energy Price

To 1985 we base this on the price of oil (in constant 1975 U.S. dollars) since oil will continue to be the fuel that dominates, and other internationally traded fuels will be priced in relation to oil. Oil will continue to be the balancing fuel. We take three alternatives:

- a *constant* real price case with no change from the January 1, 1976 price of *$11.50*** *per barrel.* This reflects the possibility that the world oil price, adjusted for inflation, remains constant to 1985.

---

* The three Technical Reports of WAES are: *Energy Demand Studies: Major Consuming Countries* (MIT Press, 1976), *Energy Supply to the Year 2000: Global and National Studies,* and *Energy Supply-Demand Integrations to the Year 2000: Global and National Studies* (MIT Press, 1977).

** All oil or energy prices are in constant 1975 U.S. dollars per barrel of Arabian light crude oil, f.o.b. Persian Gulf.

- a *rising* real price case with prices rising steadily to reach *$17.25 per barrel* by 1985. This price was chosen arbitrarily to be 50% above the constant price.
- a *falling* real price case with prices falling steadily to reach *$7.66 per barrel* by 1985. This price increased by 50% equals the constant price.

While it might seem unrealistic to assume these steady price changes, it is necessary in any study of the kind we are doing. We decided not to engage in explicit price forecasting or to try to determine exactly when prices might be raised or lowered.

For 1985-2000 we use only the constant and rising price cases. By 1985, the effect of the falling price assumption is a major one, with the low oil price helping to drive up energy demand and hold back the development of all fuel supplies. As a result, demand greatly exceeds available supplies (see discussion of 1985 cases in Chapter 8). We also change from an *oil* price to an *energy* price, as oil may no longer be the balancing source of supply. We take two energy prices for the 1985 to 2000 period:

- a *constant* energy price—equivalent to $11.50 per barrel of oil.
- a *rising* energy price—equivalent to $17.25 per barrel of oil.

2. World Economic Growth

To 1985 we make two assumptions, a *high* and a *low*. In the high case, an average growth rate of *6%* per year (1977-1985) was assumed for WOCA. We assume here a kind of upper boundary of what is likely—a world which returns to the heady economic growth and flourishing international trade of the 1960s, having resolved its many recent economic conflicts.

In the low case, an average growth rate of *3.5%* per year (1977-1985) was assumed. This is a world of the sort we have seen during 1974 and 1975, and longer in some areas. It is also a growth rate just sufficiently above population growth to allow an advance in real global GNP per capita. Significantly lower growth than this could mean some periods of stagnation or recession and some measure of political and social instability, especially in the Third World where material hopes are closely linked to world trade levels and to economic activity in the developed, industrialized nations. National economic studies done under these assumptions, when aggregated, actually result in global rates of 5.2% and 3.4% for the high and low cases, respectively, from 1977 to 1985. The detailed results of this calculation are illustrated in Chapter 2 and in more aggregated form in Figure 1-2. Here, it is important to note that we did not have time to recalculate national growth rates in light of the aggregated global rate. If we had, the results of our projections would be slightly, but we think not significantly, altered.

For 1985-2000 we reduce the growth rates slightly.

In the high case it is dropped from 6 to *5%* because it seems unlikely that the developed nations could maintain the high rates of the 1960s and 1970s into this period, with declining population growth rates.

Similarly, the low assumption shows a slight reduction on the pre-1985 period, from 3.5 to *3%* per year. This is sufficiently above the population growth estimates to allow a slight increase in the real GNP per capita in most countries. National economic studies done under these assumptions, when summed, actually result in global rates of 4.0% and 2.8% for the high and low cases, respectively, from 1985 to 2000. The detailed results of this calculation are illustrated in Chapter 2, and in more aggregated form in Figure 1-2. Again, we did not have the time to recalculate national growth rates. And again, we think our main results would have been affected only marginally if we had.

As explained above, within these global boundary assumptions, each WAES team found corresponding national high- and low-growth projections. These varied widely, as shown in Figure 1-2, with the most dramatic changes for Japan and the OPEC countries. Figure 1-3 shows the changes in GNP per capita for 1972, 1985 and 2000 over the range of WAES assumptions. Our approach is designed not to lose sight of these important national and regional differences.

3. National Policy Response

As mentioned above, we also assume public energy policy to be either vigorous or restrained to 1985, but only vigorous after 1985 and until 2000. While this is a subjective measure, it is nevertheless a key one in determining the energy requirements and supplies asso-

Figure 1-2 Regional Economic Growth Rates

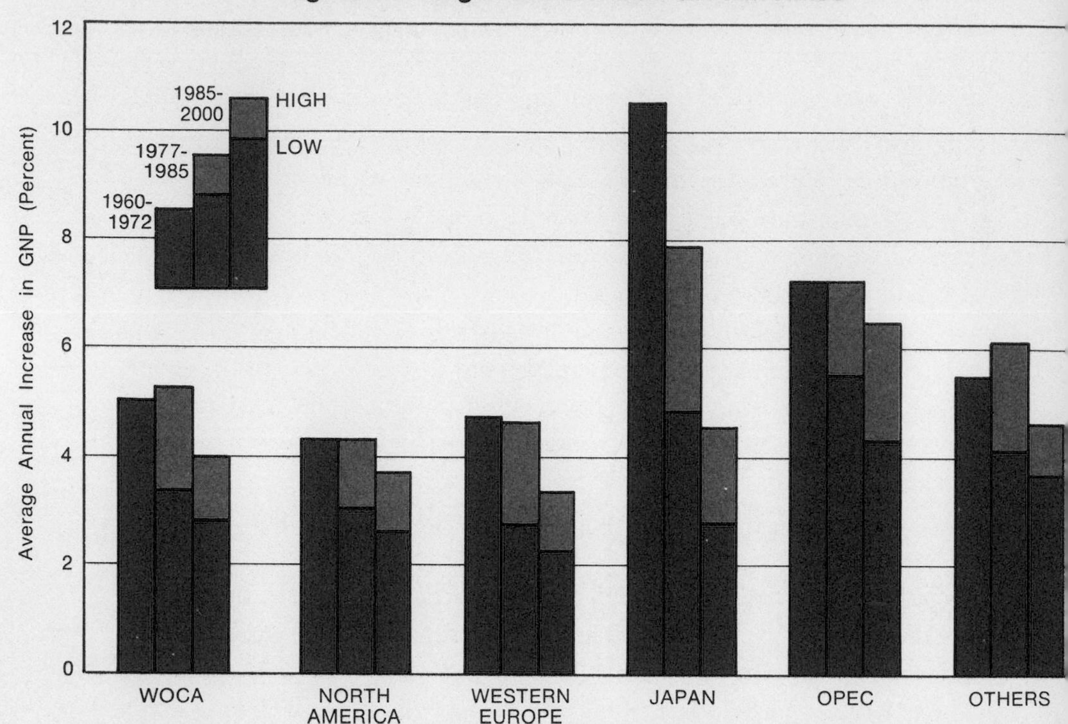

ciated with each price or economic assumption. National teams study what might be expected under these policy assumptions, examining a large number of specific supply and demand sectors. In the case of automobiles, for example, they consider present and likely technologies and their costs for improving the fuel efficiency of engines, present or impending government regulations on taxing gasoline or larger-engined cars, the probability of changeovers (possibly forced) from gasoline to more energy-efficient diesel engines, and so on. The impact of various national policy measures are reported in subsequent chapters, and in the WAES Technical Reports. As a general guideline, a vigorous* energy policy implies strong government action on both conservation and supply to the greatest degree possible, given

* A vigorous national policy response on conservation and supply includes the strongest responses that can be expected under the energy price and economic growth assumptions of the case. Market forces are assumed to act appropriately with the price and growth assumptions. Global stability is assumed. Changes in social and individual requirements may be induced by national policy responses, but these should not extend beyond what is considered politically likely in each country.

Figure 1-3 Regional GNP per Capita

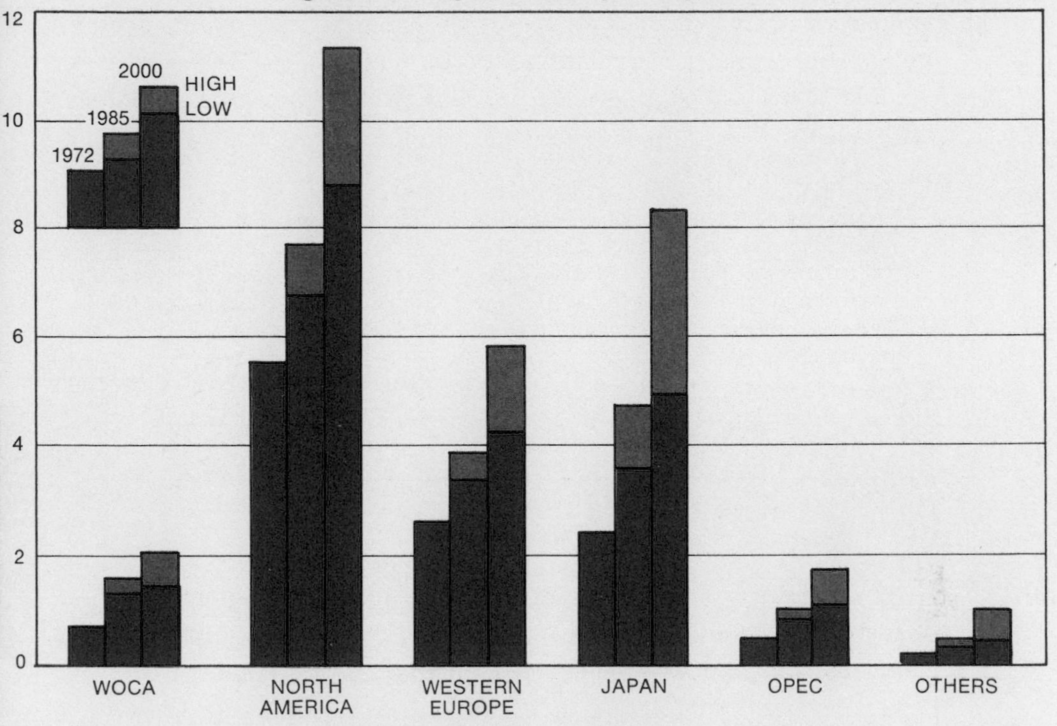

Notes to Figures 1-2 and 1-3

WOCA: World Outside Communist Areas

North America: Canada, U.S.A.

Western Europe:

WAES-Europe		plus	*non-WAES Europe*	
Denmark	The Netherlands		Austria	Luxembourg
Finland	Norway		Belgium	Portugal
France	Sweden		Greece	Spain
F.R.G.	U.K.		Iceland	Switzerland
Italy			Ireland	

OPEC (Organization of Petroleum Exporting Countries): Algeria, Ecuador, Gabon, Indonesia, Iran, Iraq, Kuwait, Libya, Nigeria, Qatar, Saudi Arabia, United Arab Emirates (Abu Dhabi, Dubai and Sharjah), Venezuela.

Others (Non-OPEC rest of WOCA): all other countries outside of Communist areas.

national political differences, under the oil price and economic assumptions of the case. On the other hand, a restrained* energy policy implies that government action on conservation and supply is the least that could be expected consistent with the price and economic assumptions.

---

* A restrained national policy response on conservation and supply is the least that can be expected under the oil price and economic growth assumptions of the case. Market forces are assumed to act appropriately with the price and growth assumptions. Global stability is assumed. Only changes in social and individual requirements induced by market forces are included.

Energy Units

Like many other measurable commodities, energy and fuels are measured by a seemingly innumerable variety of units. Attempts to achieve wide acceptance and use of a common convention—for example, the International System of Units (SI)—have met with some degree of success. But different people are accustomed to different units of measure; changes in training, habit and usage in these matters come slowly.

Not only are different systems of units in use, but there is also little consensus on precisely how to *convert* from one system to another. Given the nearly infinite variety in the qualities and characteristics of energy and fuels in the world, perhaps this is to be expected.

In this report we have generally expressed the rate of energy use (both for fuels and electricity delivered to consumers and for supply and demand of primary energy) in terms of *millions of barrels per day of oil equivalent* (MBDOE). This measure is based upon the conventional unit of a "barrel of oil equivalent" with a gross calorific value of 5.8 million British thermal units (Btu).

In many places throughout the text we have also employed other units of measurement, for example exajoules (10^{18} joules), quadrillion Btu (10^{15} Btu or "Quads") or units ap-

4. Oil Discoveries and Production Limits

We also make two sets of assumptions about world oil production. The first concerns the rate at which oil reserves are increased each year by the discovery of new fields, discoveries of additional oil in known fields, and improved recovery techniques. These *gross additions* are set as averages of 20 (*high*) or 10 (*low*) billion barrels per year (BB/yr).

The second assumption concerns limits that might be set on production by the OPEC countries, for example because of their inability to absorb further revenues. These limits are set at 40 and 45 million barrels per day (MBD), or some 10 to 15 MBD higher than in

plicable to particular energy sources, such as tons of coal or cubic meters of gas.

The calorific value of the various types and sources of coal, crude oil and natural gas varies widely. With this proviso, we give in Table 1-1 the equivalents we have used in converting from one unit of measurement to another.

To produce an electrical output of 620×10^9 kWh per year (1 MBDOE) would need power stations of 100 GWe (= 100,000 megawatts) installed capacity, given an average load factor of 70%. If the average efficiency of the power stations was 35%, one would need a fuel input of $\frac{1}{0.35}$ (= approximately 2.8) MBDOE. The difference between the input of energy as fuel and the output of energy as electricity (in this example, 1.8 MBDOE) is the transformation loss in electricity generation, which we include under the heading of "processing losses," along with the losses which occur in all other energy conversion processes.

In our discussion of primary energy supply and demand we have adopted the convention of expressing electricity from primary sources (nuclear, hydro, geothermal, etc.) in terms of the fuel input (generally expressed in MBDOE) that would be required to produce the equivalent amount of electricity output in fossil-fueled power stations.

Table 1-1 Standard Energy Conversions (Equivalent Values Lie in Vertical Columns)

Unit																		Abbreviation
Barrels per Day of Oil Equivalent[1]	—	—	—	—	—	—	—	—	—	—	0.003	0.013	0.02	1	2.7	18,000	470 million	BDOE
Tons of Oil Equivalent[2]	—	—	—	—	—	—	—	0.022	0.023	0.09	0.13	0.65	1	50	135	0.9 million	23 x 10^9	TOE
Metric Tons of Coal Equivalent[3]	—	—	—	—	—	—	—	0.034	0.036	0.14	0.21	1	1.5	76	209	1.3 million	36 x 10^9	MTCE
Barrels of Oil Equivalent[4]	—	—	—	—	—	0.0064	0.02	0.16	0.17	0.68	1	4.8	7.4	365	1 TBOE	6.4 million	170 x 10^9	BOE
Cubic Meters of "Average" Natural Gas[5]	—	—	—	0.027	0.09	1	2.7	25	27	106	155	745	1150	57,000	0.155 million	1 x 10^9	27 x 10^{12}	Nm^3 NG
Kilowatthours	—	—	—	0.3	1	11	29	280	293	1160	1700	8140	12,600	0.62 million	1.7 million	11 x 10^9	290 x 10^{12}	kWh
Cubic Feet of "Average" Natural Gas[5]	—	—	—	1	3.4	37.3	100	950	1000	4000	5800	27,800	43,000	2.1 million	5.8 million	37.3 x 10^9	1 x 10^{15}	ft^3 NG
Kilocalories	0.24	0.25	1	252	860	9400	25,200	0.24 million	0.25 million	1 million	1.5 million	7 million	10.8 million	530 million	1.5 x 10^9	9.4 x 10^{12}	250 x 10^{15}	kcal
British Thermal Units	0.95	1	4.0	1000	3400	37,300	1 therm	0.95 million	1 million	4 million	5.8 million	27.3 million	43 million	2.1 x 10^9	5.8 x 10^9	37.3 x 10^{12}	1 Q	Btu
Kilojoules	1	1.06	4.2	1055	3600	39,400	105,500	1 million	1.06 million	4.2 million	6.1 million	29.3 million	45.4 million	2.2 x 10^9	6.1 x 10^9	39.4 x 10^{12}	1.06 x 10^{18}	kJ

Calorific values are measured gross. Rounded equivalents only are given.

[1] Equivalents in other units are shown on a per annum basis.
[2] of 43 million Btu (≈ 10,000 kcal/kg net cal. val.).
[3] of 12,000 Btu/lb = 7000 kcal/kg.
[4] of 5.8 million Btu.
[5] of 1000 Btu/ft^3 or 9400 kcal/m^3.

Source: Energy Conversion Equivalents Table, Shell International Petroleum Co. Ltd.

— = insignificant
1 therm = 100,000 Btu
1 TBOE = 1000 BOE
1 Q = 10^{18} Btu
1 Quad = 10^{15} Btu

the peak OPEC production year 1973. Substantiation and further discussion of these assumptions can be found in Chapter 3.

5. Principal Replacement Fuel

One further global variable is introduced into the scenario definition for the 1985-2000 period. This is the principal replacement for oil—with coal or nuclear power as the possibilities selected. Other replacements such as oil from oil sands and shales, and renewable sources such as solar, wind and geothermal, are of course also considered (Chapter 7). Indeed, renewable energy forms represent a rapidly growing and important (if small-scale) source in many national scenarios for 2000. But their contribution was left to the analysis and judgment of national teams and did not form an overall scenario variable.

Selecting Useful Scenarios

To summarize these scenario variables:

- *for 1985*, we have three oil prices, two economic growth rates, two policy responses, and four oil production assumptions, yielding a total of 48 possible "energy futures."
- *for 2000*, we have two energy prices, two economic growth rates, four oil assumptions and two replacement fuels for a total of 32 possible energy futures.

These are more alternatives than we can or need to handle. Some of them would be highly unlikely or inconsistent; others would produce rather similar results. We wish to *learn* from the scenarios and keep to a few that seem plausible.

We therefore selected *five* useful scenarios to 1985 and couple *two* of these to the period 1985-2000. Each of these two (Cases C and D in Figure 1-4) has a coal/nuclear variant in the latter period, thus giving four scenarios for the year 2000. The choice of extending Cases C and D through to 2000 was made after the results of the projections to 1985 became clear (see Chapter 8). Case E becomes inconsistent before 1985, with demand outrunning supply. Cases A and B show that the high oil price ($17.25 by 1985) is probably higher than needed to bring on sufficient supplies to 1985. Cases C and D, with closely matched supply and demand in 1985, appeared

Figure 1-4 The WAES Scenario Cases 1972-1985-2000

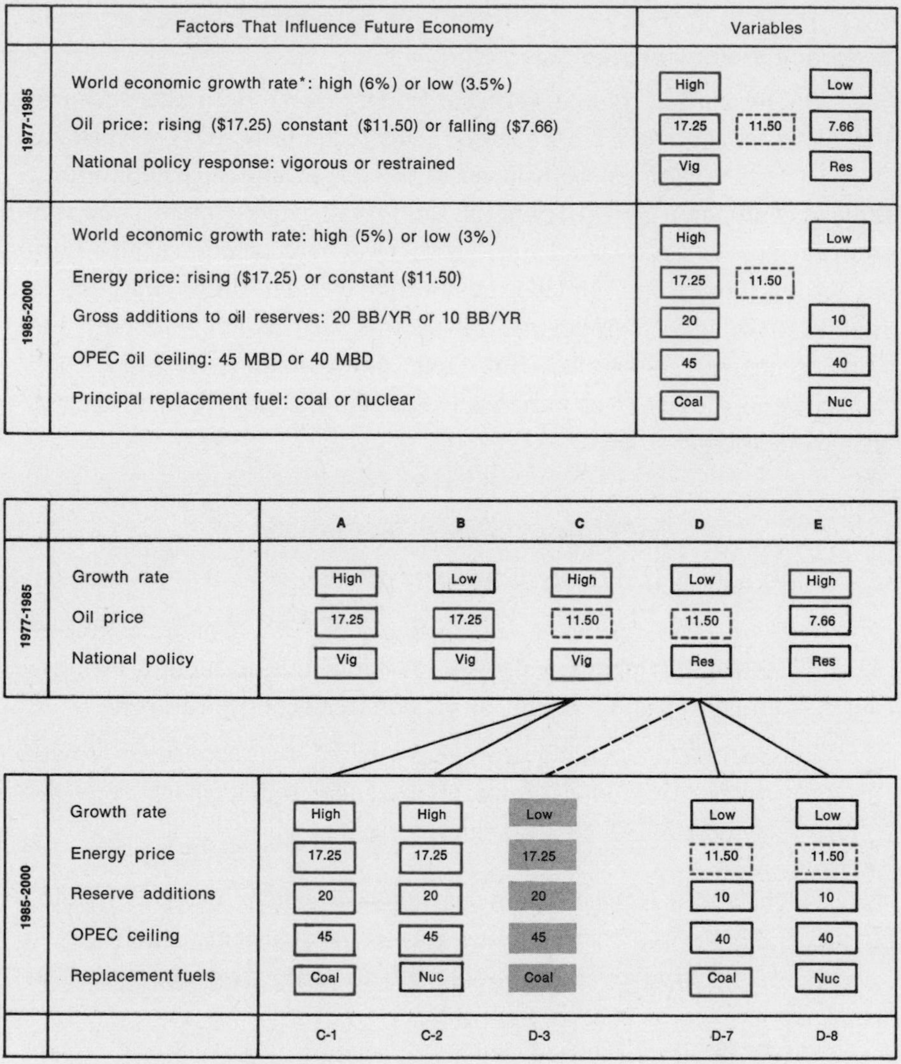

	Factors That Influence Future Economy	Variables		
1977-1985	World economic growth rate*: high (6%) or low (3.5%)	High		Low
	Oil price: rising ($17.25) constant ($11.50) or falling ($7.66)	17.25	11.50	7.66
	National policy response: vigorous or restrained	Vig		Res
1985-2000	World economic growth rate: high (5%) or low (3%)	High		Low
	Energy price: rising ($17.25) or constant ($11.50)	17.25	11.50	
	Gross additions to oil reserves: 20 BB/YR or 10 BB/YR	20		10
	OPEC oil ceiling: 45 MBD or 40 MBD	45		40
	Principal replacement fuel: coal or nuclear	Coal		Nuc

		A	B	C	D	E
1977-1985	Growth rate	High	Low	High	Low	High
	Oil price	17.25	17.25	11.50	11.50	7.66
	National policy	Vig	Vig	Vig	Res	Res

1985-2000	Growth rate	High	High	Low	Low	Low
	Energy price	17.25	17.25	17.25	11.50	11.50
	Reserve additions	20	20	20	10	10
	OPEC ceiling	45	45	45	40	40
	Replacement fuels	Coal	Nuc	Coal	Coal	Nuc
		C-1	C-2	D-3	D-7	D-8

* The period from 1973 to the end of 1975 is assumed to correspond to actual world economic conditions, with recovery to 1973 levels by the end of 1976 postulated.

National economic studies done under these assumptions, when summed, actually result in global rates of 5.2 and 3.4% to 1985 and 4.0 and 2.8% from 1985 to 2000.

We assume that the scenario variables are approximately independent of each other, within a certain range of values. Any combination of values *may*, then, be possible and no combination is automatically excluded from consideration.

on first analysis to be the most useful to study beyond 1985. Figure 1-4 shows this coupling and the 1985-2000 scenarios.

To the objection that four alternatives to the year 2000 seem too few, we would reply that the scenario cases and the projections built on them are designed for learning. We can learn enough about the basic strategic choices by using these cases together with the adaptive framework built around them. For example, all our 2000 scenarios point to possible large overall energy and oil gaps and therefore to the need for vigorous actions beyond those included in the scenario.

Furthermore, these gaps occur with "high-growth/rising-price" and "low-growth/constant-price" assumptions, and with vigorous policy responses. Clearly, a "high-growth/constant-price" scenario would produce even larger gaps—because there would be less incentive to bring on alternative energy sources—than in the "high-growth/rising-price" case. Such a scenario only makes matters worse. On the other hand, a "low-growth/rising-price" scenario might be expected to close the supply-demand gap. This "What if" possibility (Case D-3 in Figure 1-4) was considered to help us learn what kinds of adjustments will be needed to close the gaps we do foresee. In addition, it would appear that any scenario of low growth combined with our high supply assumptions might be expected to close the supply-demand gap. For a number of reasons, however, the plausibility of some such combinations could be in doubt.

It is not enough simply to select instructive scenarios. WAES also had to create a common system for making energy demand and supply studies and for doing our global balancing integrations for each of the chosen scenarios. The following "Analytical Methods" sections describe our demand, supply and supply-demand integration methods in some detail.

Analytical Methods: Energy Demand

For the demand studies we begin by dividing an economy into 69 standard sectors to span the agricultural, industrial, commercial, transport, residential and other energy-using activities in a country. These 69 economic sectors and the full results of the sector demand studies for 1972 and 1985 have already been published.* Examples of these *economic activities* are the number of tons of steel produced in a given year; value added per year in the food industry; the total

---

* *Energy Demand Studies: Major Consuming Countries,* The First Technical Report of WAES (MIT Press, November 1976).

vehicle kilometers for urban and for rural automobiles (2 sectors); and the total commercial building space requiring heating, air conditioning, lighting, and other energy-using services (5 sectors).

For each of these activities there is a technical measure of *energy-use efficiency*—how much fuel or energy is needed to do a job. These efficiencies change as technologies improve and as better devices are more widely used. The fuel needed to heat a house can be reduced by insulation and/or more fuel-efficient heating systems. For example, autos can be reduced in weight with material substitutions, resulting in the use of less fuel per distance traveled, and more people can use such energy-saving cars. The energy used in a particular activity can also increase, leading to higher energy demand. The energy intensity of agriculture, for example, would increase with additional labor-saving farm machinery.

All WAES national teams have developed base year data for 1972. Using economic analysis, econometric models, or other appropriate methods (determined by each national team), they then made projections of economic activity levels for each sector to 1985 consistent with each of the WAES scenarios (Cases A-E). Next, each national team made scenario projections for the five cases, for each sector, of the possibilities for improving efficiencies of energy use by 1985, and from this, developed a *technical coefficient* which reflects the energy use per unit of activity in each sector. These efficiencies are strongly controlled by the case assumptions, technical possibilities, lead times to develop new technologies, and many other engineering and economic variables, in addition to the "vigor" of national policy response.

The 1985 economic activity level multiplied by the technical coefficient gives the 1985 desired demand for each Case A-E. This is a slight oversimplification, but it conveys the approach.* All through this process national teams cross-checked their assumptions with national economic groups, engineers and industrial experts. Is the projection real? Does it make sense? Is it consistent?

Throughout the estimation process, the WAES national teams tried to use the best available information on environmental, technological, and economic factors that would influence future technical coefficients.

---

* The procedure we followed is an expanded and formalized version of an approach first used by the Ford Foundation Energy Policy Project in 1974.

Others are invited to check the assumptions which are made explicit in our analyses. We also encourage other energy analysts to extend our analysis of these factors.

We had to modify our methods for the 1985-2000 period, because there are more uncertainties, and we could not be so detailed. Some national teams chose to use only rather coarse measures of energy demand based on ratios such as GNP to total energy consumption and energy use per capita. These ratios have changed with time and will change in the future, reflecting shifts in economic and industrial structures, trading capabilities, energy conservation, new technologies, and so forth. Such changes were taken into account.

Some teams chose to do a more detailed job, so we developed an agreed system for condensing the original 69 sectors into 17 sectors. Others felt that it might be possible to perform year 2000 demand studies with the full 69 items; the WAES framework allows that to be done where national teams chose to do so.

Thus, by using one or another level of detail, we have made a calculation of energy demand in the year 2000 based on activity levels and energy coefficients, price and other scenario variables, and the 1985 case data. This result is energy demand for each of the four WAES cases examined for this period.

Analytical Methods: Energy Supply

Uncertainty surrounds energy supplies as well as demands. There are technical uncertainties. Can we get oil from deeper seas than now? What will be possible in synthetic fuels from coal? What will happen with nuclear reactors? How quickly can solar energy be widely and economically used? There are economic uncertainties too—especially about the true capital costs of large and complex pieces of engineering where cost overruns are habitual.

For these reasons we seek in most cases to identify the maximum *potential* supplies in each WAES country, and in non-WAES regions, in each forecast year. This means studying the wealth of data on known reserves and speculative resources. For nuclear energy we use a special method described in Chapter 6.

In our national and global studies, we assess the potential supply of each fuel using several criteria. We assess the reserve base, including expected discoveries by the year 2000. We consider, broadly, the questions of current and expected technologies, financing requirements, logistics (political, legal, institutional constraints or requirements), and environmental standards. Finally, on this basis, we estimate the maximum potential supply of each source in several selected years from 1972 to 2000. We then ask how these potential production levels might (or might not) be developed, taking account of long-term purchase contracts, prices, policy inducements and constraints, lead times to get new equipment "on the ground," and so on.

Environmental factors may be critical here. Difficult choices must be made when weighing energy/environment priorities. Air quality must be maintained, disturbed land reclaimed, waterways must be preserved—and sufficient energy must be supplied. The challenges are real ones, and are not ignored in our supply studies.

For the 1972 base year we know the facts, at least as well as they can be known. And because it takes about 10 years from decision to completion to build a nuclear plant, and from three to ten years to open a major new coal mine and build associated transport and handling systems, 1985 fuel supplies are pretty well determined by decisions already made. Oil and gas are exceptions because of the possibility, albeit small, of major discoveries of oil or gas before 1985. Also, the policies of the OPEC producers, rather than technical factors, may dominate world oil production levels. In this latter case we had to employ some special assumptions. These have already been outlined in Part I and are amplified in Chapter 3.

We could make our 1985 estimates (except for oil) with fair confidence based on these hard data simply by examining present output capacities and announced projects to expand them, making due adjustments for the economic, price and policy assumptions in each case and for particular national constraints and objectives. Then we totaled individual supplies to produce a supply profile for each nation and region and for the WOCA as a whole.

We could also use this method to the year 2000 for a few sources, such as coal and onshore oil, where the reserves and costs are fairly well-defined. In doing this we drew on the organizations of our Participants and Associates, as well as outside sources, for information and special advice. We commissioned special studies of critical fuel supply sources, such as U.S. coal infrastructure development requirements and capital costs. Chapters 3 to 6, dealing with the major fuels in turn, describe our national and global studies, our assumptions, and our findings in some detail.

But what of newer energy sources only now being developed: nuclear breeder reactors, geothermal energy, synthetic liquids or gas from coal, solar energy in various forms, wind power, and so on? Here the engineering and economic information is uncertain and often changing month by month. We include rather small "maximum technological potentials" for some of these sources in our scenarios. These estimates are, we think, in many cases conservative, and based on present technologies and costs. These sources are reviewed in Chapter 7.

Analytical Methods: Energy Supply-Demand Integration (Balancing)

Because WAES handles energy demand and supply separately, we have to compare the two, or *integrate* them. This is an essential step in our system.

When we integrate, we discover the prospective gaps (or surpluses) between national expectations for imports and probable (or possible) exports under particular scenario assumptions. Such gaps or surpluses cannot occur in the real world. A scenario that fails to balance is a signal that some assumptions require modification.

Since our purpose is to provide a framework for developing alternative energy strategies, our integration process must have multiple properties:

—It can suggest the cost, infrastructure and timing of efforts needed to balance supply and demand. What does a U.S.A.-Europe coal trade *involve*?

—It can reveal differences in a particular WAES case, such as shifts beyond market desires in the mix of fuels needed to reach a supply-demand balance.

—It can reveal the most economically efficient (least total cost to world consumers) adjustments, within tight physical and political constraints, for closing supply-demand gaps.

—It permits policy analysis: policies, costs, technical potentials, demand targets and timings can all be varied. "What if" and "What if not" cases can be tested. Choose your own case.

We have two integration procedures: 1) *unconstrained integration,* for national "preferred" fuels wherein no consideration is given to the availability of desired energy imports; and 2) *constrained integration* which forces the fuel mix to conform to global fuels availability. The major difference in the two procedures is that the first reveals supply-demand imbalances while the second attempts to balance supply and demand. The results of using these approaches is presented in Chapter 8.

Unconstrained Integrations

We start with the preferred national demands for each fuel, and the indigenous supplies for each fuel, as seen in Figure 1-5. These rarely balance, and the difference is the net imports and exports (usually imports among WAES countries). These imports and exports are calculated in national supply-demand integration studies—studies of the flows of energy from raw source (supply potential) to end-use (desired demand) and the energy losses at each stage along the way. These studies, contained in *Energy Supply-Demand Integrations to the Year 2000: Global and National Studies,** reveal national import desires and export potentials.

Integration is thus first a national summing process—adding demands by sector (transport, residential, industrial, etc.) and by kind of fuel (coal, oil, gas, electricity) in each country. This gives total "end-use" demand or energy consumed at the point of final use. To this must be added, in each country, the energy lost when raw or "primary" fuels are refined or converted and transported to the user. At the completion of this step, a national consistency check is made to insure that the results are reasonable. If they are not, the national scenario case would be revised.

The summing-up of these data into regional and global imports and exports forms the unconstrained integration. When doing this summation for non-WAES industrialized countries, we rely on published information and on Workshop members whose organizations have special expertise and operating experience in these countries. For demand estimates of non-WAES developing countries we have relied heavily on analysis by the staff of the World Bank (see

---

* MIT Press, June 1977.

Appendix I). Supply estimates for all of the non-WAES countries (especially OPEC) come from special WAES studies for each fuel.

After the summing up, we can see whether demand and supply are globally in balance—are export potentials of each fuel sufficient to satisfy import expectations? (We assume, in our approach, that potential surpluses in any region are available for export.) If yes, fuels can be allocated in world trade so that the potential exports which could enter world trade are taken up by regions with unfilled demand. Traditional export/import patterns are important here because of geography, transportation capabilities (docks, etc.), refinery sites, and many other historical factors. We use simple allocation rules depend-

Figure 1-5 Supply-Demand Balance Calculations: "Unconstrained" Integrations

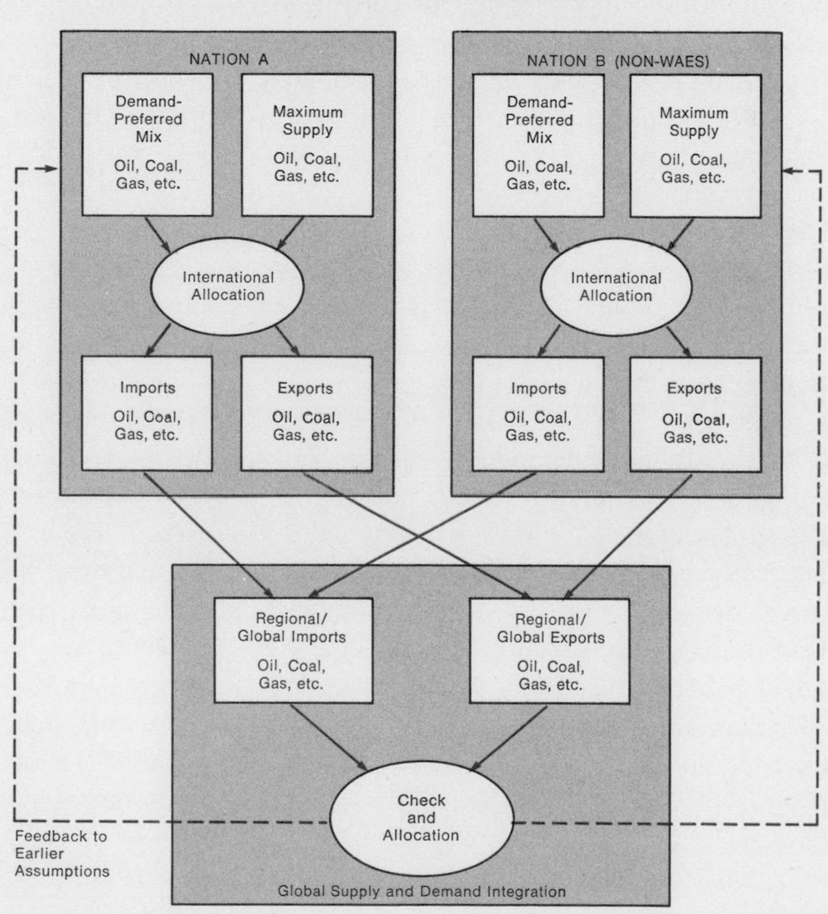

ing on these established patterns and then look at the national consequences of this sharing.

If demand and supply do not balance—if there is a shortfall of any fuel or of total energy—no world trade allocation is made. The size of each gap (or surplus) in each fuel is noted. Such gaps should be seen by nations as signals for actions—actions of the proper scale, timing, and duration to avoid or minimize the gap.

Unconstrained integrations are summations of national desires for fuels, compared to global maximum supplies. They are done to reveal prospective gaps—gaps between national expectations and the realities of globally limited fuels, under the assumptions of each case. The interaction between supply and demand is not of concern here. There is no attempt to balance supply and demand; the results of the unconstrained integrations are heavily influenced by the assumed fuel substitution rates. But the gaps revealed are highly instructive, and result directly from national preferences and expectations.

The unconstrained integrations are done for WAES by the Energy Research Group, Cavendish Laboratory, Cambridge University (England).

Constrained Integrations

In the second approach, we force shifts in the fuel mix beyond the range of consumer preferences, basing these on global supply availabilities. The unconstrained and constrained integrations use different procedures, for different purposes. They are complementary, not redundant.

Constrained integrations are done not to reveal gaps, but to see if it is possible to close them, and also to indicate details of required expansion for the energy system and related costs. The main objective of constrained integrations is to balance total energy supply and demand within maximum and minimum constraints; preferences are honored to the extent possible. The procedure requires many thousands of calculations and is done for us by WAES Associates at Atlantic Richfield Company using a specially designed, highly constrained linear programming computer model—GEMM (Global Energy Mini-Model)—with detailed inputs, constraints and adjustments provided by WAES national teams.

Our constrained integration approach incorporates all the advantages and disadvantages of linear programming applied to a dynamic environment. The results follow from use of a simplified rule for decision-making (least total cost to world consumers), which may not conform completely with reality. In addition, the results, if extended to national applications, may appear somewhat inconsistent. Nonetheless, these integrations provide a wealth of data, aggregated globally, for the potentials, constraints, and costs of closing gaps, fuel switching, and other measures.

The constrained approach starts off with the same demand and supply inputs as the unconstrained approach. The major difference lies in the global and regional uses of fuels. "Preferences" for certain fuels may not be met when global supplies (exports) are constrained. In addition, account is taken of the expansion of needed infrastructure in each region of the world, and of energy trade flows.

The procedure is shown in Figure 1-6. National, and non-WAES countries end-use energy demands are summed into regions.

Figure 1-6 Supply-Demand Balance Calculations: "Constrained" Integrations

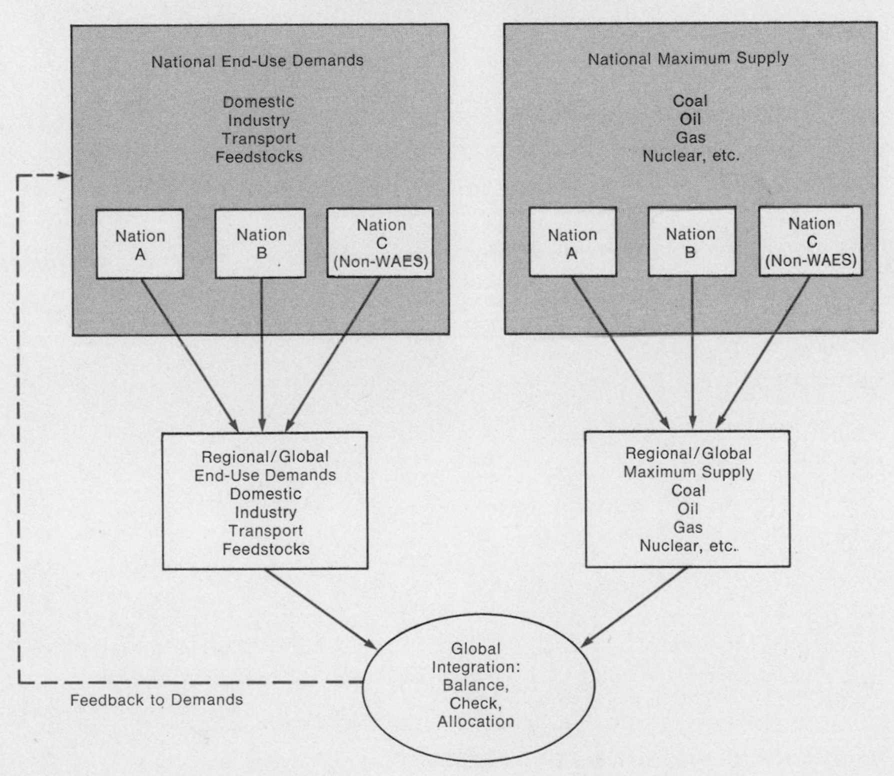

Closing the Gaps

In this report, we show how our analyses reveal prospective shortages of preferred fuels. But we take no comprehensive, detailed approach to closing such prospective gaps. Our time did not permit it. Yet several national teams did report their "responses" to this global situation. We hope many more will respond; and we have developed an approach, a method, for calculating and measuring such response. We have devised what we call energy "action programs"—packages of energy supply, or conservation, or fuel substitution that could be developed to reduce the consumption of oil and natural gas, or improve the efficiency of energy-using devices, or increase energy supply.

Each package could be added to the year 2000 scenarios without mixing it up with and upsetting our basic supply analysis —like putting a layer of icing on a cake. Each action program could be a standard size, equivalent to adding supplies or reducing demand by 100,000 barrels of oil per day (0.1 MDBOE). Thus a single deep-sea oil field (provided it can be found) with this output would form one action program; the insulation of 10 million houses to save energy equivalent to 100,000 barrels per day would be another.

One can mix them and add them to gauge their impact in closing supply-demand gaps. Some examples, with estimates of key characteristics such as capital and operating costs, final costs of energy produced or saved, lead times, technical efficiencies, etc., are included in the national supply-demand integration chapters in the WAES Technical Report on Supply-Demand Integration.* A great many additional action programs are required in order to assess the scale, scope, duration and impacts of actions required to save or add energy or to shift fuels used in a significant way. We invite, and urge, others to pursue the analysis and development of such programs.

---

* *Energy Supply-Demand Integrations to the Year 2000: Global and National Studies* (MIT Press, June 1977).

These regional demands are compared with regional supply potentials, to calculate an energy balance, with its own fuel mix, costs, and infrastructure requirements.

The calculation steps start with base-year infrastructure (the existing number of oil wells, tankers, coal mines, power plants, etc.) from which a pattern of supply, processing, transportation and capacity is calculated to meet specified 1985 or 2000 demands. In establishing these requirements, existing capacity is depreciated and new capacity is added (within a maximum potential limit). The particular pattern of supply that results from these calculations takes account of operating costs of all capacity used, and capital costs for new capacity. The pattern calculated is the one that yields least total costs to world consumers, within specified maximum and minimum constraints. Such an approach also leads to a most-efficient use of energy mix. It takes account of losses and of infrastructure expansion possibilities at each stage of processing, from resource extraction through delivery to the consumer.

The primary rule for making decisions in these calculations is: meet all end-use energy demands; make total energy supply and demand balance. To achieve this balance in stressed scenarios where some (preferred) fuels are in relative shortage while other (not preferred) fuels are in relative surplus, substantial fuel switching beyond desired mixes may be necessary. That is, the constrained integrations calculate major shifts in fuel-use patterns (when there is a shortfall of particular preferred fuels) in various regions in order that all available supplies will be used to meet end-use demands. And this fuel switching is done so that every bit of available energy content is squeezed from each and every fuel; i.e., "losses" in the processing and conversion of fuels are minimized.

It is not, in the end, the procedure followed in these global integrations that matters. It is what they show us. The integrations expose any gaps between expected or desired imports and what is likely—or potentially possible—to be available in each scenario. These are the starting points for revising perceptions of the future and for seeing what decisions and actions are needed, where and when. Some of the results of these integration procedures are spelled out in Chapter 8.

In our methodology, we can either simply report gaps and shortages and the associated policy questions. This is where the unconstrained integrations stop. Or we may try to close the gaps by forcing fuel shifts—i.e., forcing changes in preferences (as in the constrained integrations) above the maxima or below the minima—or revising national or regional assumptions. And then, we can ask: Can there be greater conservation? A shift to coal? A reduction in economic growth? And what are the costs, lead times, environmental impacts, etc., as well as the needed government actions?

We do not have the final answers to these questions. Our aim is to insure that the questions are *asked*. We hope, with our forecasting methods, to provide an effective framework in which these questions can be debated. We hope to bring clarity, focus, and a sense of urgency to the critical issues in our energy futures and their global and national implications. If we do, then our approach—and this report—will have achieved their purpose.

CHAPTER 2

ENERGY DEMAND AND CONSERVATION

Some Input Assumptions: Growth; Price; and Policy — Energy Demand Totals and Growth Rates — Energy Demand by Sector — Fuels Demand — Energy Conservation

Reducing the amount of energy consumed, especially in wasteful or inefficient uses, is widely recognized as a vital component of any successful energy strategy. Indeed, the first response of many governments to the energy crisis of 1973 was an immediate emphasis on conservation. Holding down fuel consumption is not just important—it is essential. But it is also complex.

Energy consumption is the product of innumerable decisions made by countless energy users, both large and small. Such decisions depend on a host of factors—economic factors such as incomes, costs, investments, and taxes, as well as on the price of energy. Energy use also depends on technologies and on efficiencies of energy use. It depends on climate and geography, on social patterns and norms, on government regulations, on environmental priorities and requirements, and on perceptions of the role that energy plays in our affairs.

These factors interact differently, and have different implications for different nations, and among different users. In industry, for example, energy consumption is generally based on least overall cost—given the available technologies, prices and the demand for industrial products. On the other hand, the amount of energy used to heat homes reflects climate, desired comfort, building construction standards and convenience of use—and has been generally less influenced by cost considerations than industrial energy use. The energy used in transporting people and goods has been the product of personal and institutional preferences for mobility, convenience and speed of service.

Future patterns of energy use will change as these factors change. But changes come slowly. Industrial machinery, household appliances, home heaters and other energy-using devices have long lifetimes. This causes some inertia in energy demand systems—requiring long lead times to achieve major changes in energy demand patterns.

WAES has designed an approach to estimating future energy demands that responds as accurately as seemed possible to this complexity. We incorporate the relevant factors from detailed technical and economic projections for 1985 and 2000 and break down energy use into separate energy-using activities. Examples of these *economic activities* are the number of tons of steel produced in a given year; value added per year in the food industry; the total vehicle kilometers for urban and for rural automobiles; and commercial building cubage that requires heating, air conditioning, lighting and other energy-using services.

For each of these activities there is a technical measure of *energy-use efficiency*—how much fuel or energy is needed to do a job. These efficiencies change as technologies improve and as better devices are more widely used. The fuel needed to heat a house can be reduced by insulation and/or more fuel-efficient heating systems; car engines can be made which burn less fuel for each kilometer, and more people can use such energy-saving cars. Energy use in a particular activity can also *increase*, however, leading to higher energy demand. This is often due to changing technology. The energy-intensity of agriculture, for example, would increase with additional labor-saving farm machinery and increased use of fertilizers.

To project energy demand for a given sector, one must do three things: 1) estimate the economic activity level for each sector; 2) estimate the energy efficiency measure (coefficient) for the sector; and 3) multiply the two together. This gives the energy use in the year for that economic activity.

Our demand projections done this way are of *delivered* energy —energy delivered to the final consumer in the form he uses (e.g., heating oil, gasoline, electricity). This is not the same as *primary* energy—the energy content of fuels before they are processed and converted into forms used by the consumer. (Primary fuels are coal at the mine, crude oil and natural gas at the wellhead, and so on.) Our demand projections are based on delivered energy because that allows us to consider energy-using devices and their efficiencies separately from the processes that produced and delivered that energy. We can consider automobiles and home heaters, without, *at this stage*, being concerned with oil refineries and electric power plants. It is, after all, heat in the room that is *fundamentally* what is demanded; whether such heat is produced by heating oil, or by electricity from nuclear or coal, or by solar energy, has little (if any) bearing on the level of heat—the temperature—demanded by the consumer.

Our demand projections are estimates of "desired" energy demand—energy that would be desired by the end-use consumer, given the case assumptions on energy price, economic growth and policy, if such energy were available. That is, these estimates indicate the amounts of energy that would be demanded if there were no restrictions on the availability of energy supplies (for example, oil imports, which are needed by many countries). The demand estimates are made under a variety of assumptions about alternative states of the world—including expected conservation measures. In a later stage in the WAES approach, we compare these desired demands with estimates of potential supplies. Such comparisons reveal mismatches— or prospective gaps—between expectations and availabilities. These gaps have become the focus of our work, and are described in some detail later in this report.

Our approach to energy demand and conservation is not a "normative" one. That is, we do not postulate some desirable future state and then strive to reach it. We do not choose "most desirable" conditions. Rather, we define a set of plausible alternative world situations—alternative scenarios—that we think span a range of reality in the future, and then calculate the results. All of our scenarios assume rising energy demand consistent with our assumptions about economic growth, energy prices, and changing coefficients of energy use. We do not treat explicitly the elusive question of life-styles and their impact on, or relation to, energy demand. We are not attempting to pass judgment on the relative merits, or demerits, of different future life-styles. We recognize the importance—and uncertainties—in this area, however, and invite others to pursue this topic in greater depth than we have.

This chapter summarizes the main findings of the WAES energy demand projections and our broad findings on the subject of energy conservation. The major body of work and analysis of demand and conservation was done by national teams as part of their separate national studies. It is impossible to summarize that work here. Differences among nations in this area are substantial.

Each country has its own set of national conditions, priorities and systems which can be understood only by studying each of the national projections in turn. Therefore, for the important and revealing elements of our demand and conservation work—for the detailed studies that relate the many economic, technical and social components of energy demand to their many influencing factors in each country—we refer the reader to the national chapters in two of our Technical Reports.*

We do not attempt, here, any comparisons of different national data. We have had neither the time nor the expertise, as a group, to do so. Rather, our Technical Reports provide the data for those who wish to make such comparisons.

Of course, we cannot guarantee complete consistency among our national energy demand projections. However, careful definitions, delineation and discussion of scenario variables among the national teams at frequent WAES meetings has led to a degree of confidence in the general consistency and comparability of national studies. Indeed, we feel that these data are much more nearly consistent, and lend themselves better to comparisons, than any prior estimates. While we caution that transnational comparisons can be highly subjective and potentially misleading, and must therefore be examined with care, we nonetheless invite others to study our data, make their comparisons, perform their analyses. We offer our data, in large part, so that this may be done, and we look forward to the results.

---

* *Energy Demand Studies: Major Consuming Countries* (MIT Press, November 1976), and *Energy Supply-Demand Integrations to the Year 2000: Global and National Studies* (MIT Press, June 1977).

Some Input Assumptions

Our estimates of energy demand and conservation result directly from explicit assumptions about economic growth, energy prices, and national energy policies, which we will describe here. First, however, we shall set our study in broad perspective.

From 1950 to 1972, total primary energy use in WOCA grew at an annual rate of about 4.4%. If that rate of growth were to continue (starting, say, in 1977) until the year 2000, world energy demand would then be about 230 MBDOE, or more than 10% higher than the highest of the WAES case results. For many reasons, we do not consider such historical rates of energy growth sustainable or likely to 2000. Population growth is slowing, efficiencies of use are increasing in many sectors, and energy prices—already four times 1972 price levels—are likely to continue to rise. When we make our integrations of supply and demand in Chapter 8, we discover that our economic growth rate assumptions, coupled with WAES price assumptions, result in energy demand growth that quickly outpaces plausible projections of potential supply. It follows that historically high growth rates of energy use—rates substantially higher than ours—projected into the future are simply not realistic.

In fact, because a number of important factors have changed since 1973, our projections to 1985 and 2000 do not simply assume that past trends in energy will continue unaltered. The price of oil, which had been steadily decreasing in real terms for some 20 years, suddenly rose in late 1973, along with concerns about supply availability.

Many governments have assigned high priority to energy conservation through pricing mechanisms and other energy policies. We believe such emphases on energy conservation are well placed. And we believe such emphases are—and must be—more than short-lived stop-gap measures, more than temporary political reactions to a political crisis. Conservation measures must continue, and they must increase in intensity and effectiveness. Our own studies assume a considerable measure of conservation continuing to the year 2000. More on this subject is presented later in this chapter.

Economic Growth Assumptions

In each national study, the global scenario variables of economic growth, energy price and national policy, which we discussed in the previous chapter, are translated into terms relevant to each individual country. This involves, for example, establishing high and low rates of growth of gross national product for that country to correspond to the WAES high and low global economic growth rates. This involves estimating how the composition of gross national product varies with scenarios. It also involves estimating the levels of activity of the many sectors that consume energy.

The procedure for doing this varies from country to country. Some WAES national teams made use of available econometric and other existing models. Other WAES teams relied on government experts or official government forecasts. The details are contained in *Energy Demand Studies: Major Consuming Countries.*

When this process of GNP estimating is complete, and the derived national rates are summed into global economic growth rates, the total is found to be somewhat lower than the initially assumed rates. The initially assumed annual global rates for 1977-1985 were 6 and 3.5%; the results of sums of growth rates used by WAES national teams (together with other regional rates) are 5.2 and 3.4%, respectively. The initially assumed annual global rates for 1985-2000 were 5 and 3%; the results of sums of national GNP rates are 4.0 and 2.8%, respectively.

Table 2-1 shows that annual growth rates projected for industrialized countries range from 3.1 to 4.9% over the 1977-1985 period, and from 2.5 to 3.7% from 1985-2000, compared to 4.9% during 1960-1972. OPEC countries' economic growth is assumed to return, in the 1977-1985 period, to levels comparable to its 1960-1972 rate of 7.2% per year—down from its recent 12.5% per year growth during 1972 to 1976. Finally, notice that developing countries' (excluding OPEC) growth rates (derived from a World Bank study made available to us) show continuing high growth rates although somewhat below historical rates—roughly 1% per year greater than the industrialized countries' growth throughout the 1977 to 2000 period.

Energy Price Assumptions

World oil price assumptions are also translated into national energy prices. Each nation, given different situations, will have different interpretations. Our global energy price assumptions (described in detail in Chapter 1) are three prices of oil to 1985: 1) *constant*

Table 2-1 Summary of Economic Growth Rate Assumptions by Major Region

	HISTORICAL 1960-1976					1977-1985				1985-2000			
MAJOR REGIONS	1960 GNP	1960-72 GROWTH RATE	1972 GNP	1972-76 GROWTH RATE	1976 GNP	1977-85 G.R. HIGH	1985 GNP HIGH	1977-85 G.R. LOW	1985 GNP LOW	1985-2000 G.R. HIGH	2000 GNP HIGH	1985-2000 G.R. LOW	2000 GNP LOW
North America	764	4.3	1264	1.6	1345	4.3	1970	3.0	1760	3.7	3388	2.6	2576
Western Europe	533	4.7	924	2.0	1003	4.6	1500	2.7	1278	3.3	2425	2.2	1774
Japan	75	10.5	248	3.3	284	7.9	563	4.8	434	4.5	1091	2.7	646
Australia & New Zealand	26	4.5	46	3.6	53	5.0	82	3.6	73	4.3	155	2.8	111
INDUSTRIALIZED COUNTRIES	1398	4.9	2482	2.0	2685	4.9	4115	3.1	3545	3.7	7059	2.5	5107
Lower Income Countries	61	3.7	94	2.3	103	4.4	152	2.8	132	3.1	240	2.5	191
Middle and Upper Income Countries	158	6.2	325	5.9	409	6.6	727	4.5	608	4.9	1490	3.9	1079
ALL NON-OPEC DEVELOPING COUNTRIES	219	5.6	419	5.1	512	6.2	879	4.2	740	4.6	1730	3.7	1270
OPEC	34	7.2	78	12.5	125	7.2	235	5.5	202	6.5	604	4.3	380
TOTAL WOCA	1651	5.0	2979	2.8	3322	5.2	5229	3.4	4487	4.0	9393	2.8	6757

All GNP values are given in billions (10^9) of constant 1972 U.S. dollars.

All growth rates are given as percent per year increases in constant dollar value of GNP.

All data (except historical) for North America, W. Europe, and Japan result from WAES national team estimates, based on WAES global economic scenario assumptions. All data for other regions result from special studies, performed by the World Bank, based on WAES global economic scenario assumptions. Data for 1976 are preliminary.

Developing countries' regions are defined in Appendix I. Other regions are defined in the notes on Figure 1-2 of Chapter 1.

($11.50*); 2) *rising* (to $17.25 by 1985); and 3) *falling* (to $7.66 by 1985); and two energy prices to 2000: *constant* ($11.50) and *rising* (to $17.25 by 2000).

We recognize that energy demand is sensitive, in varying degrees (in different countries, in different sectors, among different groups, etc.), to energy prices. Energy prices have been, and are, a patchwork resulting from different costs, policies, needs and conditions in different countries, for different fuels. Governments intervene to modify market prices through direct controls, taxes for revenue, and allocations. This maze of prices and pricing policies makes generalizations on the subject of oil or energy prices nearly impossible.

To try to understand the extent of the current diversity of national fuel prices, we conducted an informal survey among the WAES national teams. Figure 2-1 shows the results of this compilation. The large differences—among fuels, among countries and among different users in each country—in energy prices are apparent.

The data in Figure 2-1 should be considered as preliminary and approximate. They result from rather quickly compiled surveys from a variety of information sources. They represent averages for fuels which differ in energy content and, in some cases, in definition, from country to country. These differences place great uncertainties on any detailed comparisons of such data.

Figure 2-1 suggests that coherent and consistent energy-pricing policies must be developed if there are to be viable strategies on the energy scene in the last quarter of the 20th century.

Clearly, energy prices are central to studies of future energy alternatives and of future energy demands. However, in our demand studies, we were not able to deal as thoroughly with this complex subject and in as much detail as might be desirable. We invite others to probe this important topic in depth. For our purposes, we assumed that up to the year 2000, energy prices would *tend* to reflect world oil prices. That is, we adopted the admittedly simplifying assumption that fuel prices would tend to be based on the heat content of each fuel (using oil as the standard for pricing), modified by the relative utility (ease of transport, cleanliness of use, etc.) of each fuel. Each

---

* Prices are given in constant 1975 U.S. dollars per barrel of Arabian light crude oil, f.o.b. Persian Gulf.

Figure 2-1 Selected Energy Prices Comparison — Mid- to Late-1975

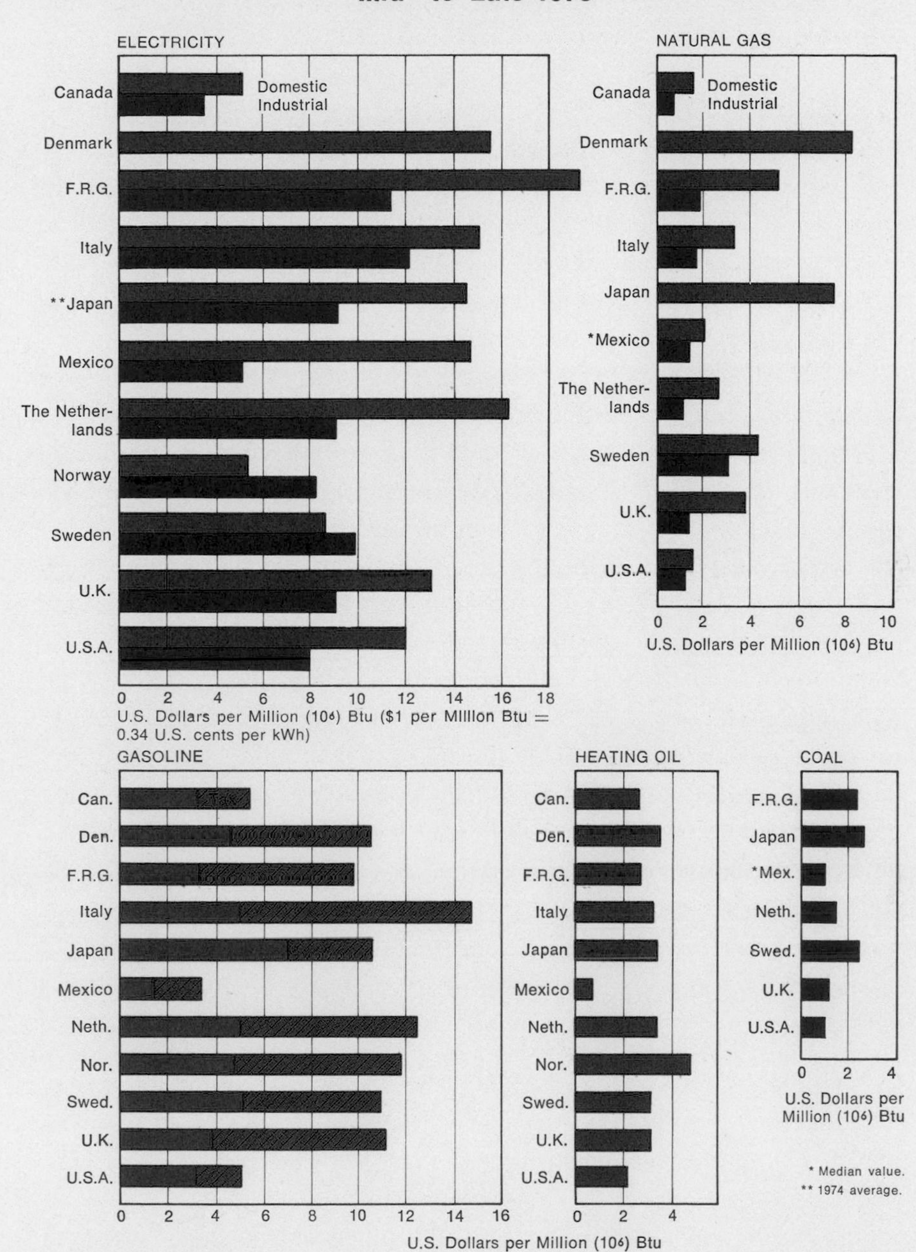

national team interpreted the relative rate and mechanisms of such a tendency in ways appropriate to each country.

We believe this assumption to be valid over the *long term*

that we are studying. In the shorter term, of course, the differences in fuel prices can have profound effects. However, we believe our simplifying assumptions to be useful and necessary here.

Energy Policy Assumptions

We specify *vigorous* and *restrained* energy policies to 1985, with only vigorous policies after 1985. On the demand side, these terms refer to vigorous or restrained government actions designed to improve energy-use efficiencies—to promote conservation—without affecting the economic assumptions of each case. Such policies differ markedly, of course, from country to country.

Like energy prices, current energy conservation policies are a patchwork, reflecting different styles, priorities, and needs in different countries. A mix of taxes, regulations, controls, public education programs, cajolings, and incentives combine to make up government energy policies. We recognize the substantial impact of such policies on energy demand. Without trying to define what each country should do or how (if at all) national policies should be consistent with one another, we specify two policy levels, two extremes in a range of plausible government approaches. By postulating both vigorous and restrained policies, we intend to span the range of plausible future policy options. Definitions of these policy types can be found in Chapter 1; specific national vigorous and restrained policies can be found in the national chapters of the First and Third WAES Technical Reports.* A further general discussion of conservation policies occurs later in this chapter.

Energy Demand Totals and Growth Rates

If the assumptions in our demand projections are correct, then growth in energy demand will slow down. In particular, we find that as a function of growth in GWP (Gross World Product), growth in energy demand *can* decline—if only slightly—over the last 25 years of this century. The ratio of energy growth to GWP growth would be about 0.82 to 0.87 from 1977 to 2000, given our range of assumptions, compared to the recent (1965-72) ratio of 1.02.

---

* *Energy Demand Studies: Major Consuming Countries* (MIT Press, November 1976), and *Energy Supply-Demand Integrations to the Year 2000: Global and National Studies* (MIT Press, June 1977).

If we look only at the industrialized areas of Western Europe, Japan and North America, however, we find a more significant decline in the energy/GNP ratio than for the WOCA average. This is because our studies indicate relatively greater energy conservation potential in the industrialized countries than in developing countries.

The changing relationship between energy and economic growth is a central observation of the demand projections which result from our assumptions about economic growth rates, energy prices, and government policies. Higher energy prices and increased emphasis on energy conservation will reduce consumption, compared to past rates of use. The relationship between economic development and energy growth is an important and complex topic. It is one that warrants further study by others.

Details about our projections of the economic-energy linkages for industrialized countries are contained in the national chapters of the First and Third WAES Technical Reports; for developing countries this information can be found in Appendix I of this report.

Table 2-2 shows the aggregate results of our demand studies. The first two rows—fossil fuel and electricity *end-use* demands—add up, in each column, to "Delivered (End-Use) Energy Demand." This is the energy delivered to the end-use consumer, in the form he uses it. This is the main product of our energy demand studies. The next row, "Processing Losses," is the result of supply-demand integrations (see Chapter 8). Processing losses represent the energy that is lost when primary fuels (e.g., crude oil, coal, nuclear) are converted into end-use forms (e.g., gasoline, electricity). Primary energy demand is demand for primary fuels; it is the *total* energy demanded in any year, including processing losses.

Table 2-2 shows that delivered energy demand in the year 2000 is higher in the high economic growth cases, even though these cases have a rising energy price. Our demand projections to year 2000 represent average annual growth rates in energy use in WOCA of 2.6 to 3.5% from 1972, compared to a 4.4% growth from 1950 to 1972. Desired electricity demand in our cases grows at 3.8 to 5.1% per year globally, compared to recent historical rates of 7.5%.

Our projections of demand for electricity are "preferred" demands. That is, they represent estimates of the amount of electricity in the fuel mix that would be desired, *if* preferred quantities of oil, gas and other fuels were available for import at the case price and other

Table 2-2 Energy Demand Totals for the World Outside Communist Areas — 1972, 1985, 2000

(all numbers are in MBDOE)

	1972	*1985*		*2000*			
		C	*D*	*C1*	*C2*	*D7*	*D8*
Economic Growth	—	High	Low	High	High	Low	Low
Energy Price	—	Const.	Const.	Rising	Rising	Const.	Const.
Policy Response	—	Vig.	Res.	Vig.	Vig.	Vig.	Vig.
Principal Replacement Fuel	—	—	—	Coal	Nuclear	Coal	Nuclear
Fossil Fuel Demand	50.8	76.2	70.6	117.6	116.7	94.9	93.3
Electricity Demand	5.8	11.0	10.2	20.1	22.9	16.2	18.3
Delivered (End-Use) Energy Demand	56.6	87.2	80.7	137.7	139.5	111.0	111.6
Processing Losses (incl. international bunkers)*	23.6	36.0	33.4	60.2	67.2	48.8	53.9
Primary Energy Demand (See Fig. 2-2)	80.2	123.2	114.1	197.9	206.8	159.9	165.4

Note: Columns may not add precisely, due to rounding.

* The energy used in these categories is calculated in the supply-demand integration process (see Chapter 8).

assumptions. Under these *unconstrained* conditions, we project electricity demands to grow more slowly than in the past. Yet when imports are *constrained* by global supply potentials, we find that even these levels of electricity demand cannot be met (see Chapter 8).

Electricity is a clean, convenient and important fuel. Our projections of electricity demand are more than triple today's levels. For certain uses—e.g., lighting, home appliances, industrial power, telecommunications—there are virtually no substitutes for electricity. For other uses, fossil fuels or solar energy may be economical substitutes.

Within electricity generation, substantial energy savings can be achieved by the use of power plants' waste heat in district heating systems. Similarly, the cogeneration of electricity and process heat in industry has large potential for greater use of available energy. While several national teams included such programs in their estimates, we have made no comprehensive studies of either district heating or cogeneration. This remains an important task for others.

All of our scenarios result in increasing energy consumption per capita worldwide. Even in the lowest energy consumption per capita case (D-7), it exceeds 1972 levels by 12% in year 2000. Among the major industrialized regions of the world, energy/capita exceeds 1972 values by 50 to 80%. For developing countries, the ratio grows by 80 to 150%. There will naturally be variations from country to country.

An important component of this projection is, of course, our population growth assumptions. We assumed that population grows, in all cases, at the rate of the U.N. Medium projection.* This rate, a global average of 2.2% per year, gives a population of about 4.8 billion in the year 2000 for WOCA.

The range of total primary energy demands in the WAES cases is also shown in Figure 2-2. We have noted that the highest case is well below the level that would be reached by simple extrapolation of historical rates. Yet Figure 2-2 illustrates that none of our cases shows a decline in energy use, or even a severe flattening of the demand curve. Our projections show continued but moderated growth in energy demand. According to WAES studies, total primary energy demand for WOCA in the year 2000 ranges from 160 to 207 MBDOE —a spread of about 25%.

Some Important Factors

There are abundant reasons for different demand levels and rates of growth in our projections. Factors that affect demand vary among nations: each country has its own mix of energy policies, economic growth plans, life-styles, geography, etc., to add to the worldwide demand picture. For example, the projected energy demand growth rates for most WAES countries are considerably lower than historical rates—reflecting significant improvements in energy-use efficiencies. On the other hand, demand growth in many other countries, particularly developing countries, is near or above historical averages.

---

* *Concise Report on the World Population Situation in 1970-1975 and Its Long Range Implications,* United Nations, New York, 1974.

Figure 2-2 A Range of Demand Futures for the WAES Cases

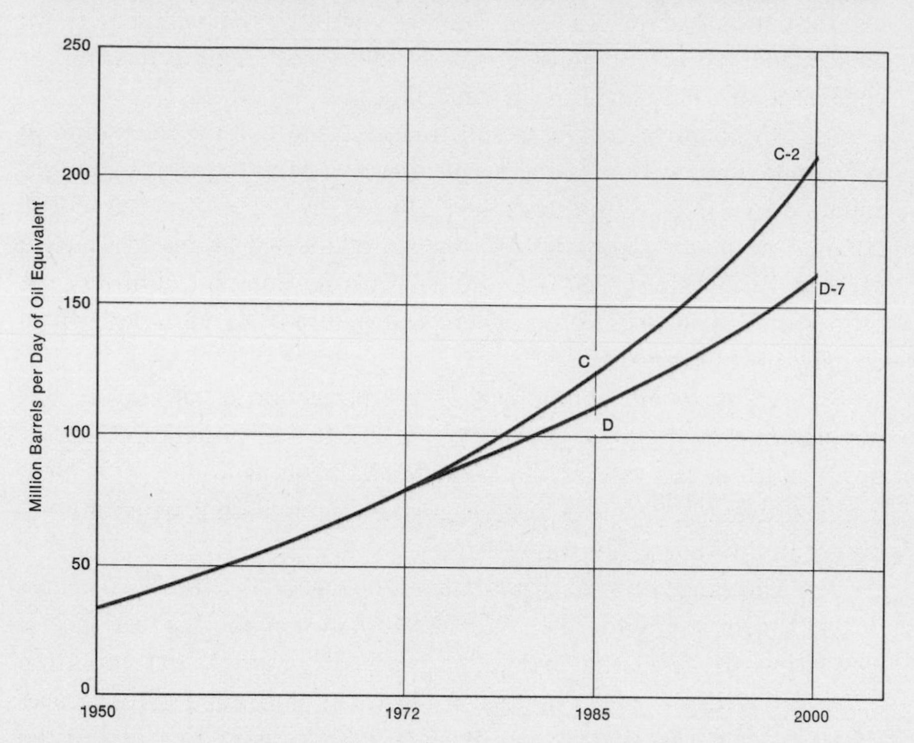

The degree of structural changes in economies assumed in our projections varies from country to country; substantial structural shifts could, potentially, have even greater impacts than technical improvements.

Naturally, some countries show greater potential for conservation than others. Some countries expect to increase energy-intensive activities, others expect to reduce such activities. Global generalizations are difficult. Yet some have argued that energy growth rates (in the industrialized countries, especially) are nearing "saturation"—that is, that the industrialized world is nearing the limits of its capacity to use more and more energy.

We are not convinced that that is generally the case, at least not before the year 2000. Nonetheless, growth in energy consumption in selected categories may well be nearing "saturation," as population growth rates decline and living standards tend to stabilize. Figure 2-3, adapted from the Danish chapter in *Energy Demand Studies*, is an example of expected saturations in some energy uses. The situation in certain other WAES countries is not unlike that in Denmark: energy

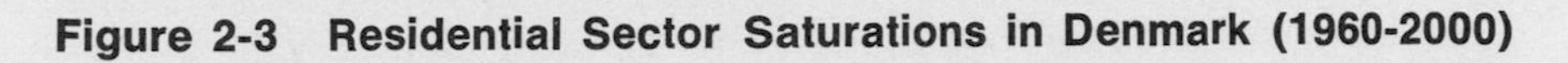

Figure 2-3 Residential Sector Saturations in Denmark (1960-2000)

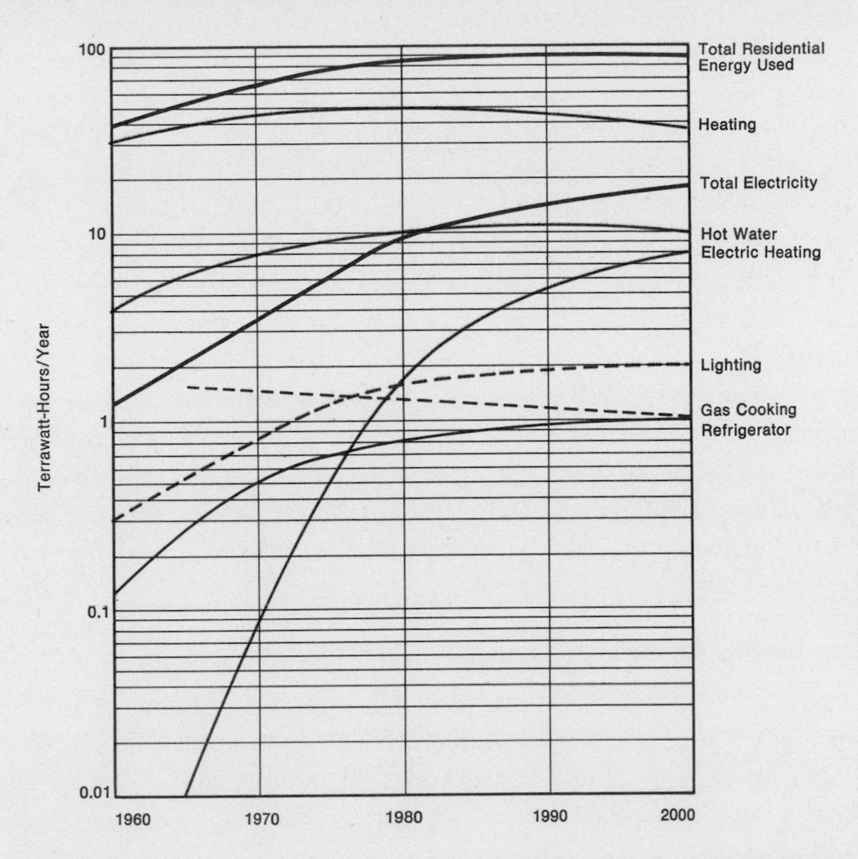

use by each appliance starts on an exponential growth curve but flattens out as 100% saturation is neared. Like Denmark, many WAES countries have flattening energy-use curves for such residential uses as refrigerators, stoves, hot water heaters, washing machines and televisions. Other appliances, of course, are still on rising curves: freezers, dishwashers, clothes dryers—not to mention unknown "new uses." Yet, it is difficult to imagine a great many energy-intensive "new uses" in the residential sector turning up in the next 25 years to alter our projections.

Is Zero Energy Growth Possible?

A seemingly logical extension of the effects of saturations in energy-use categories might appear to be zero growth in total primary energy use. However, we believe that zero energy growth (ZEnG) would be extremely difficult to achieve, worldwide, by the year 2000

—much more difficult than is commonly thought. (ZEnG, here, means a *gradual* reduction in the annual rate of increase in energy consumption so that, by the year 2000, the annual growth rate in energy use would be zero.) The WAES cases to 1985 and 2000 include estimates of substantial actions to reduce energy consumption. Yet our cases show demand in year 2000 to be well above a level that would result from ZEnG.

Some countries could possibly approach ZEnG by year 2000. For others, the severe conservation measures and the required structural and institutional changes in the economy would be unprecedented.

Our "prospective shortages" of energy argue for even stronger measures to reduce demand than we have assumed in our scenarios. But, our assessment of the efforts required to save energy, and the costs (in reduced economic growth) of drastically lowering demand, lead us to conclude that ZEnG is a more realistic target for *after* the turn of the century.

Developing Countries' Demand

There are, as we have indicated, important differences for energy demand between industrialized countries and developing countries. We have so far concentrated our attention mostly on the results for industrialized nations—since WAES national studies are mostly of these countries.

But we have also estimated energy demand for developing countries, following a somewhat different procedure. With over 90 countries in this category, by the World Bank's classifications, and with little current data available, only a highly aggregated study could be attempted. In addition, much of the energy consumed in these countries is local and noncommercial—firewood, cow dung, and vegetable waste. In India, for example, noncommercial energy has been estimated to constitute nearly 60% of total energy consumption in 1960 and 48% in 1970, and it is expected to remain a significant, though decreasing, percentage.* WAES studies of demand in the developing countries focused on commercial energy because statistics are available only for commercial fuels. While local, noncommercial

* *Report of The Fuel Policy Committee,* Government of India, 1975.

sources are important, and much of the world depends on these fuels for their energy, it is nearly impossible to find usable data on them.

We have relied extensively on the World Bank for projections of future economic growth rates and estimates of the income elasticities of energy demand for the developing countries for the period 1975-2000, based on the WAES scenario assumptions. From these factors, we can obtain energy demand projections. A summary of the supply and demand analysis for these countries can be found in Appendix I.

Non-OPEC developing countries were grouped according to level of income as generally done in World Bank studies. *Low-income countries* are those with annual per capita income below $200 (in 1972 U.S. dollars). These are South Asian and low-income African countries. The second category used by the World Bank is *middle-income countries*—with annual per capita income over $200. This group includes countries in East Asia, Central and South America, middle-income Africa and West Asia. OPEC countries are treated separately.

During the period 1960-1972, the developing nations more than doubled their consumption of commercial energy and increased their demand for electric power by 250%. In 1972, the developing countries accounted for around 15% of total WOCA energy consumption. Total commercial energy consumption within the developing world in 1972 was slightly less than 10 MBDOE. Of this total, 25% was in lower-income countries, 58% in middle-income and 17% in OPEC countries.

Energy demands have been estimated by first projecting economic growth rates for the developing countries, consistent with the WAES scenario cases. Economic growth rates by region are estimated through the World Bank SIMLINK model.*

WAES high and low economic growth rate assumptions for industrialized countries are used as input to this model, which then provides projections for the major developing country regions. Once economic growth rates are produced, the energy required to sustain those levels of economic growth is estimated. This was done by first

* "The SIMLINK Model of Trade and Growth for the Developing World," World Bank Staff Paper #220, October 1975. Also in *European Economic Review*, Vol. 7 (1976) pp. 239-255.

examining the historical relationship between regional economic growth and energy consumption. Since we assume that real energy prices will either remain constant or increase by 50% by year 2000, depending on the WAES scenario, the historical income elasticity of energy demand has been revised downwards for the period 1976-2000.

The analysis shows that non-OPEC developing countries as a group are expected to increase their commercial energy consumption between four and five times their 1972 level by year 2000. This is based on an average economic growth rate of 4 to 5% per year until 2000. OPEC countries are expected to achieve a level five to eight times their 1972 level of energy consumption by the year 2000, based on a real economic growth rate of 6 to 7% per year during 1972-2000. Figure 2-4 shows energy consumption projections for developing countries by groups—OPEC, low-income countries, and middle-income countries.

Primary energy consumption in the developing countries during 1972 constituted approximately 15% of total WOCA consump-

Figure 2-4 Primary Energy Consumption of Developing Countries

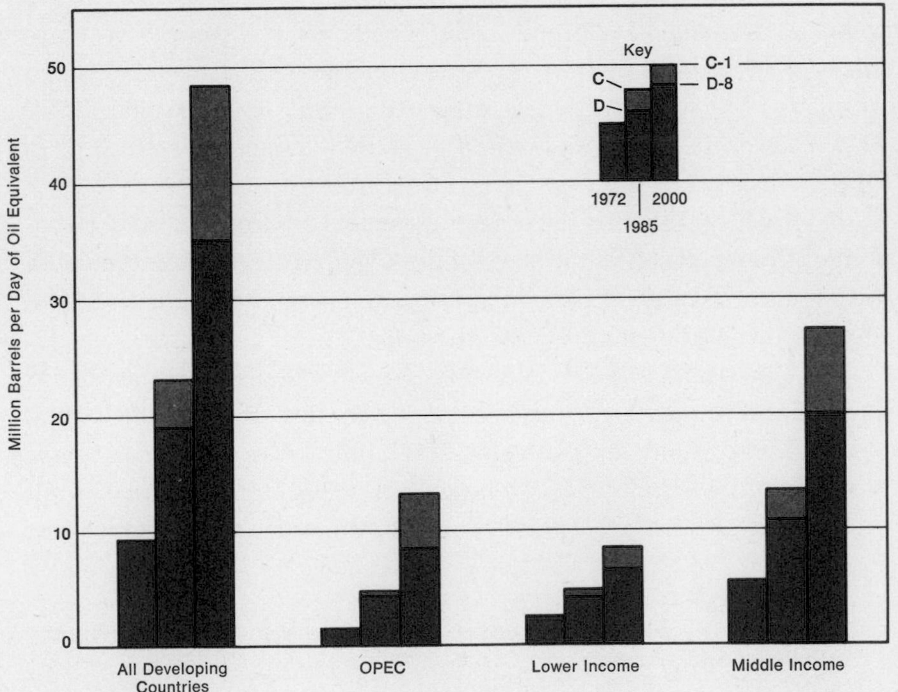

tion. This share will increase significantly as these countries become more industrialized. WAES global demand studies suggest that these countries could consume as much as 25% of WOCA energy by year 2000, as Figure 2-5 illustrates.

Figure 2-5 Shares of WOCA Energy Consumption

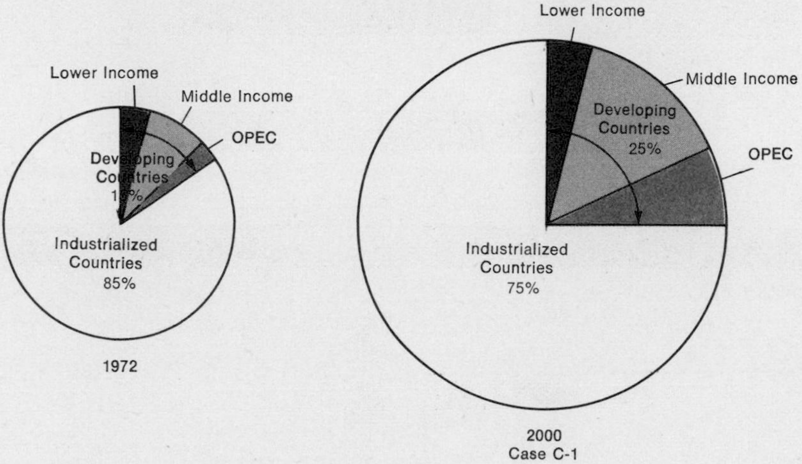

Energy Demand by Sector

Throughout our studies of energy demand in WOCA, the relative amounts of energy used by various sectors of the economy can vary considerably from country to country, even within the largely industrialized WAES group. This should come as no surprise—we have repeatedly stressed the importance of national differences in considerations of energy demand and conservation. Nonetheless, we offer here some broad generalizations, based on the globally aggregated data shown in Figure 2-6.

Energy demand for transportation grows from 15 MBDOE in 1972 to between 29 and 36 MBDOE in 2000. But transportation's share of total delivered energy declines from about 27% in 1972 to about 25% by 2000. Of course, this worldwide result does not apply to every country or every region. The trend is seen especially in North America, where transportation accounts for about 26% of end-use energy in 2000, down from 32% in 1972. Also, the U.S.A. projections show an absolute reduction in gasoline use, from 4.5 MBDOE in 1972 to 3.8 MBDOE in 2000, due to improvements in fuel

Figure 2-6 Energy Uses by Sector, WOCA

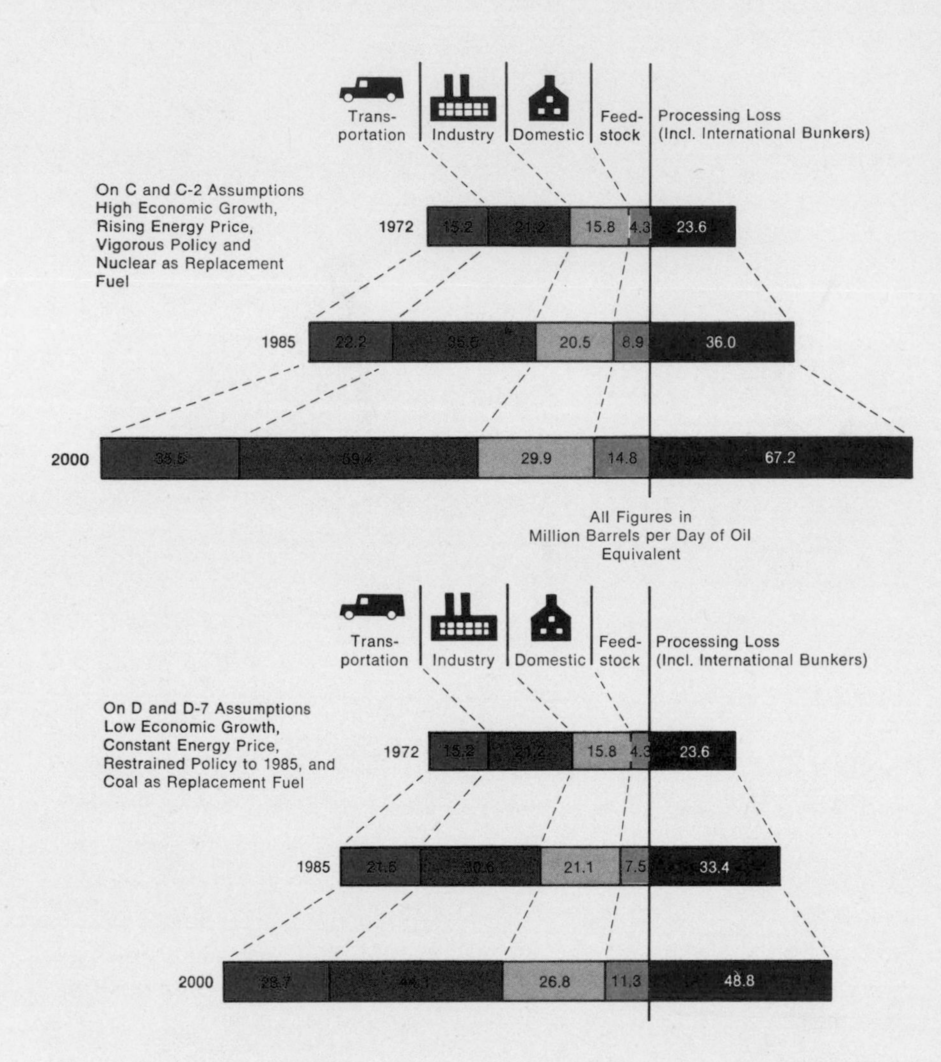

economy, and increased use of diesel engines. The trend is also seen to a lesser extent in Western Europe, where there are projected shifts toward more use of public transport. In developing countries, industrial growth overshadows growth in energy used in transportation.

This is a great simplification, for many other factors influence demand patterns in each region and country. Nonetheless, for the transportation sector we can generally conclude that:

—All modes of transport—passenger and freight—will achieve various improvements in efficiency;

—growth in freight and air transport will continue to be high; and

—projected efficiency improvement will reduce the auto's share of the transport sector's end-use demand.

In all regions (except Japan, where a shift *away* from energy-intensive industries is projected), there is a substantial increase in the relative share of energy use in industry—especially fuels used for petrochemical feedstocks. Industry's share of total energy use grows from 37% in 1972 to between 40 and 44% by the year 2000. Growth in petrochemicals is even greater, so that fuels used in this industry in the year 2000 are projected to amount to around 11% of total end-use energy—up from just over 7% in 1972. This is due, largely, to anticipated petrochemical industry growth and limited potential for improving energy efficiency. Thus, all industries and petrochemicals account for as much as 53% of delivered energy demand in 2000. Their 1972 share was 45%.

Growth of energy demand in the residential and commercial sectors ("domestic" energy use) in WOCA is reduced because of lower population growth rates, saturations in some energy-intensive uses, and improved building standards (better insulation). This trend, which is evident in the global aggregates in Figure 2-6, is especially true for a few largely industrialized countries. On the other hand, in certain European countries, in Japan, and in many developing countries, increasing size of dwellings, introduction on a wide scale of modern appliances, and growth in many service sectors result in just the opposite effect—domestic sector energy use *increasing* as a share of the total.

Figure 2-6 also shows increasing "processing losses" as a share of primary energy. In 1972, processing losses (not including international bunkers) amounted to 26% of total primary energy demand. Our projections show rather striking growth in these losses—as high as 61.8 MBDOE by year 2000, or over 30% of total primary energy demand.

This is due largely to projected growth in electricity generation at 35% efficiency, and small, but increasing, amounts of synthetic fossil fuels produced at an overall efficiency of about 60%. Such increases in energy losses have profound implications for fuels strategies.

When certain fuels are in relatively short supply, it is desirable, of course, to reduce "losses" of these fuels in processing.

Our analyses of the trade-offs and potentials in reducing such losses are presented in the results of the "constrained" supply-demand integrations (Chapter 8).

In sum,

—The fastest-growing demand sector is petrochemical feedstocks; industrial energy use also grows faster than the average.

—In the transport and domestic sectors, energy demand grows slower than the average.

—While there are noticeable differences, there is also considerable similarities in the distributions by sector of energy demand among major regions.

—Industry remains by far the largest demand sector.

Fuels Demand

Our demand projections for individual fuels are determined, in each case, from observed trends and certain built-in elements of resistance to change in the mix of fuels in various markets. Figure 2-7 shows "preferences" in WOCA for individual fuels. The scenario cases were selected to show the widest difference between WAES cases in terms of fuel preferences. As can be seen, the cases do not differ greatly: consumer preferences for different fuels are about the same for all scenario cases to 2000.

However, some interesting general observations can be made. Demand preferences for oil and gas, which together accounted for about 74% of all fuels used in 1972, amount to around 60% of fuels demanded by 2000. Clearly, these are and will remain, relatively "preferred" fuels. Of course, these are our "unconstrained" desired fuel demands. We discover, in supply-demand integration, that such demand levels cannot be met from available supplies in the year 2000. Nonetheless, these demand expectations give useful information. They tell us what the preferences are, and what changes in the fuel mix will have to be made.

Nuclear and hydroelectric demands grow rapidly, owing largely to the projected nuclear expansion programs (Chapter 6) in our cases. These sources of electricity grow by 600 to 900% from 1972,

Figure 2-7 Primary Energy Demand by Fuel Type, WOCA

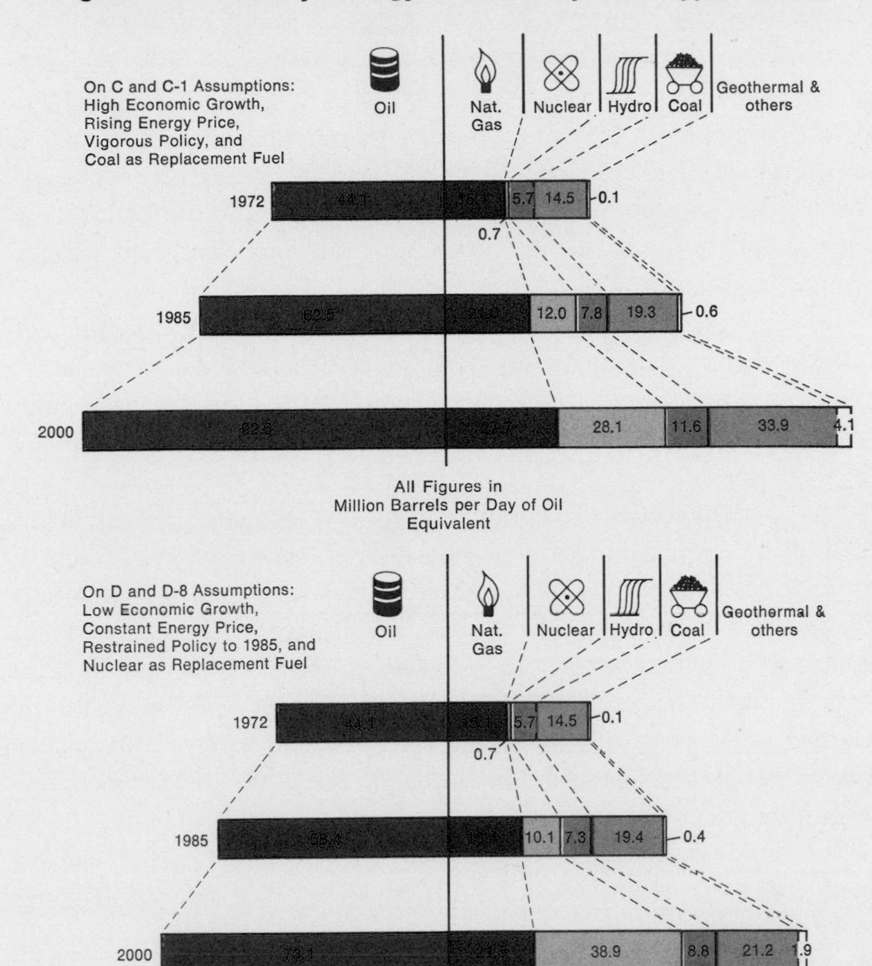

so that by 2000 they account for 20 to 30% of total primary energy demand. Preferences for coal change little as a percentage of the total—from 18% in 1972 to 13 to 17% in 2000—although desired coal demand reaches over 34 MBDOE by 2000, from 14.5 in 1972. Coal has not been and is not projected to be, a "preferred" fuel. This, in spite of our observation (Chapter 5) that coal may well be a fuel in relative surplus by the end of the century.

Further details about components of and constraints on desired demand for each fuel are contained in the various fuel chapters—3 through 7.

Priority Uses of Oil

It is important to look carefully at oil demand because, as we shall see in more detail later, we find major prospective shortages of oil. Our projections, as Figure 2-7 shows, reveal continuing expectations for large amounts of oil. Although declining somewhat as a percentage of total energy demand, WOCA increases its demand for oil from 44 MBD in 1972 to between 73 and 93 MBD by 2000 in our cases. Future desires for petroleum, thus, are nearly double the size of today's uses.

Yet, as we have noted, there will not be sufficient supplies of oil, given our assumptions, to meet these demands. To what extent are we really dependent on oil? To what extent can we substitute other fuels for oil? Figure 2-8 shows the makeup of the global desired demands for oil.

Transportation and "non-energy use" (mostly petrochemical feedstocks) comprise a growing portion of future oil uses (up to 50%). For these purposes, liquid hydrocarbons are all but essential. There has been little evidence to date of any potential for large-scale switching of automobiles from gasoline or diesel fuel to other fuels (e.g., methanol or part-methanol, or electricity) or of using coal on a massive scale as a petrochemical feedstock. A rather sizable shift toward public transportation (including electric) is already included in our studies.

The automobile is responsible for 60% of transport sector energy demand today; autos consume some 12% of all primary energy, and over 20% of all oil. We project the auto's share to drop to about 40% of all transport uses, and about 15% of all oil uses by 2000. Yet the magnitude of the oil used in automobiles will be large in 2000—nearly 15 MBD by our projections. Clearly, the automobile is an important user of oil, and will continue to be.

In short, transport and petrochemical uses may well be "noble uses" of petroleum. For a long time to come, we shall be dependent on oil for these uses.

The situation is not the same for the other uses of oil. Industry consumes about 20% of total preferred future oil demand in our projections. The potentials for shifting industrial uses (mostly process heat) to coal are substantial. With concerted effort, oil use in industry could be substantially reduced.

The same is true of the domestic sector. For space heating,

Figure 2-8 Oil Uses, WOCA

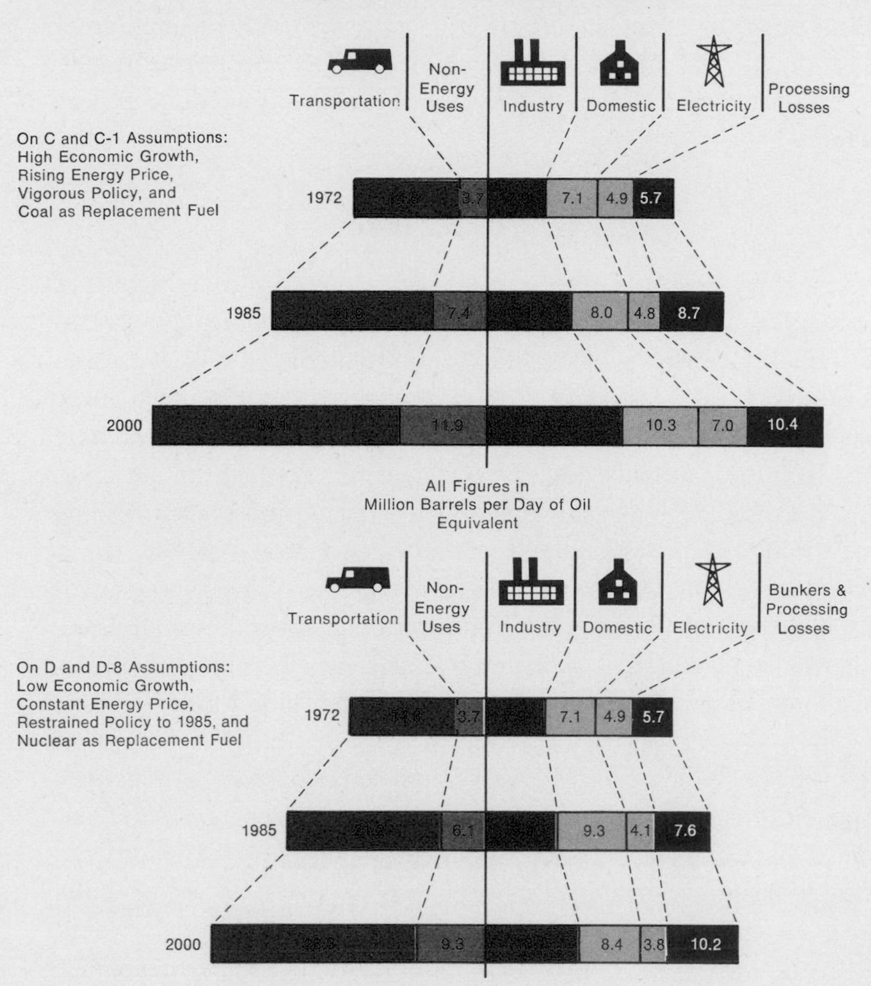

water heating, and cooling needs, heat pumps and coal (conceivably, given the proper burning device) and solar thermal energy could certainly replace oil. Oil is not essential for domestic uses and can be replaced over time by other energy sources.

Nor is oil essential in large quantities for electricity generation—although it seems likely to be a valuable source in small quantities in gas turbines for peak electric power generation.

Our conclusion is that oil is essential for just two major uses: transportation (private auto, airplanes, ships, trucks) and petrochemical feedstocks. Oil will be needed well beyond the year 2000 for these purposes. The implications are clear: energy-saving actions should be directed at *saving oil* or at *shifting uses* away from oil to some other energy source. The need for actions aimed in this direction is urgent. It is, or should be, a high priority among efforts to reduce energy demand.

Energy Conservation

Closely coupled to energy demand projections are estimates of the potentials for saving energy—the potentials for energy "conservation." Energy conservation is vitally important to our energy future. It is a topic whose opportunities and challenges deserve much more careful analysis, much more detailed probing, than we have been able to do. Here, we can only describe our general approach to conservation, summarize our main findings on the subject, and highlight some of the specific measures assumed in the WAES national studies.

People everywhere are calling for greater "energy conservation." Yet the term carries a thousand meanings for a thousand users. It vaguely connotes thrift, economy, efficiency. In our studies, we have chosen to adopt a relatively restricted definition of energy conservation.

In our studies, energy conservation refers only to those actions and policies that increase the technical efficiency of energy use. Conservation here does not involve significant changes in the traditional growth of economic activities, changes in life-styles, or major shifts away from energy-intensive activities, other than those that would result regardless of our overall economic assumptions. We recognize the difficulty of strictly separating technical efficiencies from economic activities (for example, separating improvements in automobile kilometers/liter from changes in people's driving habits). Yet the attempt to decouple these factors, as a first-order approximation, enables us to be more clear and explicit about conservation measures and expected energy-savings effects.

Energy demand reduction, in a wider sense, can include the effects of structural change in the economy, product mix changes, changes in trade patterns, and changes in social patterns and norms. While several national studies treat one or more of these areas in a particular national context, we make no attempt to summarize or generalize these concepts here.

Approach and Background

The procedures adopted in the Workshop call for separate estimates of activity levels (e.g., size of and distances traveled by the auto fleet) and energy efficiencies (e.g., gasoline consumed per distance traveled) for various sectors of the economy. To provide such estimates requires that consideration be given to the capital stock turnover in each of these sectors (e.g., number of years required to replace the stock of autos). Details of how this was done in the separate national studies can be found in the WAES Technical Reports.

To estimate energy conservation effects, estimates of the energy required per unit of activity are calculated based on the price and policy assumptions of a particular case. Wherever possible, such calculations are related to the introduction, at a specific rate, of specific technologies that are economically justified and perform the same service with less energy at the WAES case price of energy.

Reduced rates of growth of energy demand in the WAES cases are due in large part to efficiency improvements. Such changes do not happen automatically. Behind them are rising energy prices, and sustained and vigorous actions by governments. Investments, planned replacement of equipment, and intelligent legislation and regulatory action are required. Lead times are significant because of the delays in changing energy-using equipment and consumer behavior. Indeed, the time required for full implementation of conservation measures can, in some cases, be very long because of the long lifetimes of many energy-using devices.

Lead times can often be longer, for example, than required for full implementation of supply expansion measures. On the other hand, conservation measures—unlike supply expansion programs—are at least partially effective almost at once. A *portion* of the capital stock turns over immediately, bringing small but gradually increasing savings. With many supply expansion measures, which take years to bring to production, *no* gain is realized until production begins.

A variety of conservation and other energy-saving measures have been adopted by various countries since 1973, including fiscal measures (for example, taxes, tax relief, loans and grants), regulations and standards (for lighting, heating, insulation), encouragement of action by common means (public transportation, total energy systems), control (import controls and allocations), public education programs, and research and development. An excellent summary of recent measures adopted by many developed countries is furnished in "Energy Conservation in the International Energy Agency," a 1976 review by the Organization for Economic Cooperation and Development (OECD).

Energy Conservation in Sectors

Energy conservation in transportation has been assumed in various degrees by WAES national teams in their studies.

Conservation measures in this sector include, generally, improvement in the efficiency of transportation equipment; switching from a less efficient to a more efficient mode; and switching away from overwhelming dependence on oil.

Throughout the various national studies, automobiles have been carefully examined for potential oil savings beyond those conservation measures already assumed in the estimates. With few exceptions automobiles are built by private industry. Yet they must conform to government standards or tax schedules which affect safety, environmental effects, and fuel economy. Implementation of programs to improve automobile efficiency thus usually involves responses to market conditions (consumer tastes, or fuel prices) and/or changes in the tax or regulatory structure.

Within WAES, the largest improvements are estimated for autos in North America. Europe and Japan for many years have had tax structures for fuel and automobiles which, coupled with geographical conditions and personal preferences, have resulted generally in smaller, more efficient vehicle fleets.

Our estimates of average fuel efficiency for the auto fleet in Canada and the U.S.A. are 20 and 21.7 miles per gallon (U.S.) in 1985, respectively, with the average in the U.S.A. ranging from 27 to 29 mpg by 2000. Many other national studies assume little or no auto efficiency improvements. For their own situations, holding to no worsening of auto efficiency may itself be an indication of relatively strong conservation actions.

Energy conservation in industry is greatly influenced by energy price policies and tax treatment of investment. This is because, in industry, investments in energy conservation measures are made by professional economic decision-makers accustomed to making project evaluations sensitive to energy price expectations. This is also true of public or semipublic enterprises—even though they may be responsive to noneconomic criteria to a greater degree than private firms.

Nearly all WAES national studies project substantial efficiency improvements in industrial processes. In some areas, 1% per year improvements in efficiency to 2000 are estimated. Japan's already relatively efficient iron and steel industry is projected to reduce its energy use per unit of value added by as much as 15% from present levels by 2000. Thus, industrial energy conservation, which is assumed to be more price sensitive than other sectors, and which typically requires 20 to 30 years for turnover of major energy-using equipment, also plays a major role in our scenario projections.

Energy savings in buildings are achieved in response to both price increases and government policy. Some of these changes involve conventional technologies such as increased insulation, while in other cases new technologies such as heat pump applications are expected.

Throughout the scenario cases, assessments are made of the trade-offs for the buildings sector between added capital investment, which reduces energy requirements, and the higher fuel costs accompanying the rising prices of energy. Owing to the long periods of time required to replace any nation's stock of buildings (up to a century or more to replace the *entire* stock) and the very large fraction of total energy consumed by buildings (about 25%), improvements in this sector are typically the result of long-term programs, bringing substantial savings only at high levels of implementation. In addition, improvements of existing structures can be as important as better design of new ones, because the stock turns over so slowly.

Several special problems occur in organizing energy conservation for buildings. In contrast, say, to the industrial decision-maker, the builder of houses and commercial space aims for the lowest first cost, and not for the lowest lifetime costs, which include the cost of energy to heat, cool and light the building. In addition, information about such life costs is not usually available to him or to the buyer. Thus, in an environment of rising prices, investment in energy conservation is likely to be below the optimum level.

Actions by governments to alleviate problems of limited capital availability or government regulations concerning building design can contribute to energy savings. Of course, in countries where housing is largely owned or subsidized by governments, there is a need for the government to act as a responsible owner and lifetime operator and to implement economically justified conservation measures. Heat pumps, if widely installed and used, afford great potential for reduced energy use in buildings. Several national studies include projections of the distribution of heat pumps, their assumed efficiencies (called coefficients of performance) and consequent energy use. However, we have done no comprehensive, overall global study of the potentials and challenges of widespread heat pump use. We hope others will.

Our studies of the residential and commercial sectors show wide variations from country to country in the projected improvements. In some European countries and in Japan, increasing size of dwellings, even if they are well-insulated, tend to offset the gains from improved design and better efficiency. In the U.S.A., on the other hand, better residential dwelling insulation on new and existing structures is projected to improve efficiency by as much as 40% by the year 2000—bringing this efficiency measure to a value just slightly better than today's average European standard. Also, Scandinavian countries, with already highly efficient buildings, estimate improvements as high as 20 or 30% as a result of more building insulation by 2000. The conservation potentials in buildings are substantial. Our studies have not ignored them.

Implementation of Conservation Measures

Implementation of energy conservation technologies has a decision-making character quite unlike that of supply technologies. One of the reasons is that conservation equipment typically comes in small units—one house, one car, one industrial heat exchanger, etc. This means that they can be implemented on a decentralized basis—with the decision made by an individual building owner or plant manager without the wide public approval necessary for a large supply plant. Of course, a corollary of this is that many such small decisions, rather than a single large one, must be made in order to save a certain amount of energy.

A few energy-saving measures can be implemented in a relatively centralized way. The use of waste heat from power plants through district heating, and the cogeneration of heat and electricity in industry, are examples. While the success of such schemes has been proven in only a few countries, their potential for reducing overall demand is high.

Energy conservation is generally not limited by technology,

but is limited by economic factors. For example, consumer preferences for home insulation and more efficient automobiles and appliances are influenced by energy prices, and not by technology. Even heat pumps involve relatively simple and known technologies. This is a significant advantage in view of the technical and economic uncertainties surrounding many energy options.

Energy conservation measures are usually helpful to environmental quality. With some exceptions, conservation measures reduce environmental impacts by reducing the impacts of the production, processing and use of energy.

The conservation measures in our studies will not occur without concentrated action by many groups in all countries. Decisions and actions are required by governments, industries and individual consumers if such energy-conserving measures are to be implemented. Energy conservation is dispersed and requires widespread adoption.

But energy conservation is essential. Conservation may be our best, cheapest, most accessible alternative energy source. It unquestionably must play a central role in global and national energy strategies to the end of the 20th century and beyond.

CHAPTER 3

OIL

Producing Oil — Factors That Determine Oil Supply — Proven Reserves at the End of 1975 — Additions to Proven Oil Reserves 1950-1975 — Future Gross Additions to Reserves — WAES Assumptions for Gross Additions to Reserves — Oil Demand — Coupling of Oil Demand and Oil Discovery Rates — The Real World — OPEC and Non-OPEC — Non-OPEC Oil Production 1975-2000 — Non-OPEC Production 2000-2025 — OPEC Production 1975-2000 — OPEC Production 2000-2025 — Total World Outside Communist Areas' Oil Production — World Outside Communist Areas' Oil Supply/Demand Balance in 1985 and 2000 — A More Optimistic Future?

Oil is now the major fuel in the World outside Communist Areas (WOCA). It accounts for over half of the energy supply in many countries. Therefore, an analysis of the future supply of oil is critical in any long-term energy study.

Our analysis identifies the following principal factors that will affect future oil production: known reserves; the rate at which new discoveries or improved production techniques will add to them; the level of oil demand and the rate of production that OPEC countries might allow. Alternative assumptions are made about each of these factors, and oil production profiles that meet oil demand for as long as possible are developed.

The main conclusion is that if all of the oil-producing countries of WOCA allow production to increase to meet demand, limited only by technical factors, oil production will peak and decline before the end of the century. This holds true over a wide range of assumptions. However, an even more important limitation on production is a governmental one—the rate at which OPEC countries are willing to produce. If recent government announcements of possible production limits take effect, production could reach a peak early in the 1980's and stay at that lower level until the end of the century and beyond. Peak production could be delayed until the late 1980's if OPEC production limits are higher. Even in this case, oil production would at best remain on a plateau through the 1990's.

Oil production beyond 2000 is also projected. The assumptions used beyond 2000 are highly speculative, and the production profiles should be viewed as such. Oil production figures for the period 2001-2025 are not used elsewhere in the WAES energy analysis because our time frame is 1975-2000.

Producing Oil

Some background information and terms used in the oil industry are described here before proceeding with the analysis.

It is a popular misconception that oil is found in vast underground pools or lakes that, once discovered, can be pumped dry with relative ease. Nothing could be further from the truth. Oil is found trapped in the small spaces or pores between individual rock grains—rather like water in a sponge. Over time oil and water seep through porous rocks until impervious rock is reached. Figure 3-1 shows one example of an oil field. The porous rock is capped by an impervious layer of rock or cap rock. If this cap rock is in the form of a dome, as shown in the figure, then fluids, mainly water but often including oil, gradually accumulate under it. Oil gradually separates from the water and finds its way to the top of the structure, accumulating in the porous rock under the cap rock and above the layer of water in the porous rock. This is the oil field or reservoir. Gas is also often found in the reservoir, either dissolved in the oil or as a gas cap above the oil as shown in Figure 3-1.

The existence of underground structures which might contain oil can be determined from the surface by seismic techniques. Although these have improved in the past 20 years, it is still necessary to drill a well to determine whether a structure actually contains oil in commercial quantities.

Figure 3-1 The Geology of Oil

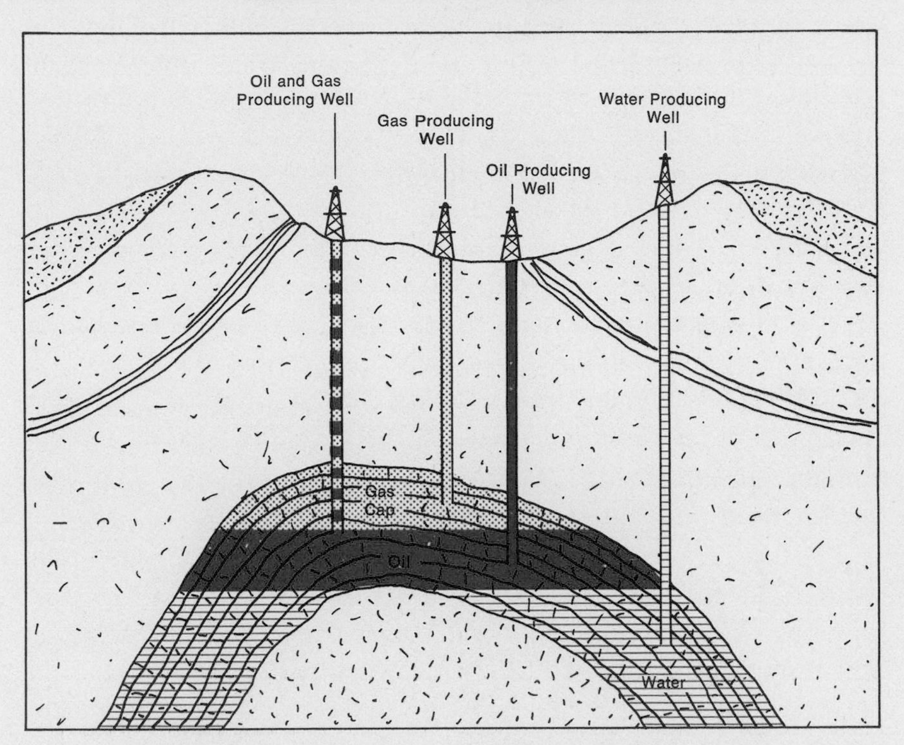

Drilling a well into an oil-bearing structure releases the natural pressure in the reservoir, forcing oil into the well. Oil produced by this natural pressure is known as *primary* production. The portion of oil in the reservoir that can be produced in this way varies from field to field. It depends on such factors as the porosity of the rock and the viscosity of the oil. Faults in the rock structure may also affect primary production. It is difficult to state a global recovery average for primary production, but in the U.S.A. for example, it yields an average of about 25% of the oil in place.

The amount of oil recovered can be enhanced by pumping water or gas into the reservoir to increase or maintain the reservoir pressure. Such techniques are known as *secondary recovery*, and their success varies. Some fields respond well because of their physical characteristics. Others do not. In the U.S.A. secondary recovery is mainly responsible for increasing recovery from 25% of the oil in place in the 1940's to about 32% by 1975. In some Middle East countries, such as Iran where large reserves of natural gas are available, injection of gas into oil fields to improve recovery has already begun.

A further method of improving recovery rates from oil fields is to lower the viscosity of the oil so that it flows more easily through the pores of the rock. This can be done by heating the oil for example, by injecting steam, or by injecting chemicals to dilute the oil. This is called *tertiary recovery*. It is not widely used yet since the technology is costly and not well-developed. In tertiary recovery, energy is injected into the oil field in the form of heat or chemicals, causing a reduction in the net energy recovered in the oil by the amount used in the heat or chemicals.

The major share of the world's oil production comes from conventional oil fields of the types just described. There are other current, potential sources of oil such as synthetic oil from coal, production from nonconventional sources, and natural gas liquids.

In 1976, coal liquefaction only provided a significant amount of oil in South Africa. Even then supplies represented only a tiny proportion of world oil supply. The future potential of coal liquefaction is discussed in Chapter 5. Nonconventional alternative sources of oil, for example, oil sands, heavy oil and oil shale are discussed in Chapter 7. Currently they supply less than 0.2 MBD.

Natural gas liquids (NGL) are by-products of natural gas production. They generally include hydrocarbons that can be extracted in liquid form from natural gas when it is produced. They are generally blended with crude oil and its products and contribute to oil supply rather than to natural gas supply. In 1975, around 5% of WOCA oil production was from NGL. In this analysis, NGL will be implicitly included with oil production, and whenever oil supply or production is mentioned, it is taken to mean crude oil from conventional sources plus NGL.

Proven reserves of oil are usually defined as oil that is recoverable from known reserves with today's technology and prices. Therefore, in addition to primary recovery, proven reserves include potential production based on using secondary and tertiary techniques—where such techniques have been evaluated and are expected to be used in the fields.

Ultimately recoverable reserves are an estimate of how much oil will eventually be produced. They usually include new discoveries plus an allowance for enhanced recovery as secondary recovery becomes more widely used and as tertiary recovery techniques are developed.

Factors That Determine Oil Supply

Estimates of ultimately recoverable reserves of crude oil for the total world have been made by many oil geologists. Table 3-1 shows how estimates have increased from around 500 billion barrels in the early 1940's to about 2,000 billion barrels in the 1960's. Since 1960, estimates by geologists have tended to converge to around 2,000 billion barrels, although some estimates are well above this figure. While this figure might owe something to its convenience, it is unlikely that this agreement of oil geologists is coincidental.

It therefore seems reasonable to take 2,000 billion barrels as one estimate for the ultimately recoverable reserves of oil from conventional sources and NGL. Some geologists are less optimistic, and we therefore also looked at a lower figure of 1,600 billion barrels. Finally, at the end of this chapter we look at a more optimistic case where ultimately recoverable reserves were assumed to be 3,000 billion barrels.

These estimates may include reserves in the deep sea and Antarctica. Such reserves will only be produced at high cost and

Table 3-1 Estimates of Total World Ultimately Recoverable Reserves of Crude Oil for Conventional Sources

Year	*Source*	*In Billion Barrels*
1942	Pratt, Weeks & Stebinger	600
1946	Duce	400
1946	Pogue	555
1948	Weeks	610
1949	Levorsen	1500
1949	Weeks	1010
1953	MacNaughton	1000
1956	Hubbert	1250
1958	Weeks	1500
1959	Weeks	2000
1965	Hendricks (USGS)	2480
1967	Ryman (Esso)	2090
1968	Shell	1800
1968	Weeks	2200
1969	Hubbert	1350-2100
1970	Moody (Mobil)	1800
1971	Warman (BP)	1200-2000
1971	Weeks	2290
1975	Moody & Geiger	2000

with advanced technology and are unlikely to provide significant volumes of oil before the end of the century.

These estimates are for the total world, including Communist countries. The WAES analysis is for WOCA. Therefore, we have to subtract estimated reserves in the Communist area. Geologists do not agree on oil reserves in the Communist World, but a figure of 20% of the total world reserves will be assumed for this analysis. Using this percentage for all our cases gives estimates of ultimately recoverable reserves in WOCA of 1,300, 1,600 and 2,400 billion barrels.

However, these numbers are in one sense academic. Oil can only be produced from fields that have been discovered and for which production facilities have been installed. Oil production is based on proven reserves, the rate at which they are added to, and on the rate at which production facilities are developed. The importance of ultimately recoverable reserves is that they determine how long a rate of additions to reserves can be maintained.

Each field has a *potential* production rate that depends on the size of the field, its geological characteristics and its installed facilities, e.g., secondary recovery facilities. In some areas, governmental controls on production may also be an important factor. Together, these factors set an upper limit to the annual production from an oil field.

Primary recovery relies on natural pressure within the reservoir and the maximum yield is obtained by releasing this pressure gradually. In general this means that it is impossible to produce more than 10% of the recoverable reserves in any one year without reducing the amount of oil that can be eventually recovered. In some fields, it may be possible to produce at a faster rate. In others, the rate may be lower, but overall, a proven reserves-to-production (R/P) ratio of 10 to 1 is probably the minimum feasible for the world's oil reserves.

Applied to the world however, an R/P ratio of 10 to 1 would imply that all known oil fields are producing at the maximum rate. As discoveries continue to be made, some fields will be under development. So, although they are included in the world's proven oil reserves, it will be some years before they produce any oil. This was the position in 1976 of the Prudhoe Bay field in Alaska and many North Sea fields. Thus an R/P ratio of 15 to 1 is probably a more justifiable estimate of the maximum rate of production from the aggregate of the world's proven oil reserves. As discoveries decline, the R/P ratio may decrease toward 10 to 1.

These are factors that form the technical basis of the WAES estimates of future oil production. The level of proven oil reserves for each year is determined by adding to the proven reserves at the end of one year the gross additions to the reserves during the year and subtracting the *actual* production during that year. The maximum *potential* production from these reserves is determined by the annual proven reserves and the limiting R/P ratio. As long as oil demand is less than this *potential* production, *actual* production will equal demand. When demand exceeds *potential* production, demand will be limited to this potential, which becomes *actual* production.

Thus, it is possible to project an oil supply profile based on the following factors:

1. An estimate of proven reserves in a base year;
2. An estimate of future annual rate of gross additions to reserves;
3. An assumption of a limiting R/P ratio that, for the reasons discussed above, is assumed to be 15 to 1; and
4. An oil demand curve.

Proven Reserves at The End of 1975

Estimate of proven oil reserves at the end of each year are published by journals such as *Oil and Gas Journal* and *World Oil*. These estimates are made by asking oil companies and governments of oil-producing countries to give estimates of reserves. The estimates are stated to be reserves recoverable at current prices and with current technology. Therefore they probably include oil recoverable by primary production plus oil recoverable by secondary or tertiary recovery—where the potential has been evaluated and facilities are planned.

Estimating recoverable reserves is subject to much uncertainty and can therefore change—up or down—from year to year as new assessments are made. In this analysis, we have used the *Oil and Gas Journal* figures published at the end of 1975.* The figures are

---

* Since this paper was written these estimates were revised downward by 40 billion barrels at the end of 1976, after taking into account 1976 production.

broken down by region in Table 3-2. Remaining proven reserves in the world at the end of 1975 were estimated at 658 billion barrels, of which 555 billion barrels were in WOCA. The uncertainty in this figure is illustrated by comparing these estimates with those given by *World Oil* for the end of 1975—579 billion barrels, of which 500 billion barrels are in WOCA.

Table 3-2 Total World — Reserves and Cumulative Production to the End of 1975

	Remaining proven reserves (billion barrels)	*Cumulative production to end of 1975 (billion barrels)*
OPEC: Saudi Arabia	152	23
Other Middle East	208	61
Other OPEC	90	55
Total OPEC	450	139
North America	40	133
Western Europe	25	2
Rest of the WOCA	40	17
Total Non-OPEC	105	152
Total WOCA	555	291
Communist Countries	103	50
Total World	658	341

(Source: *Oil and Gas Journal*, 29 December 1975)

Table 3-2 shows that about 80% of the remaining proven reserves are in OPEC countries, with the Middle East especially important. Saudi Arabia is in a class by itself with 152 billion barrels—about 27% of the proven oil reserves in WOCA. Much of the difference between *World Oil* and *Oil and Gas Journal* estimates is found in the latter's higher estimates for proven reserves in Saudi Arabia. The non-OPEC countries in WOCA have only 105 billion barrels distributed fairly evenly between North America, Western Europe and the rest of WOCA.

Table 3-2 also shows oil production to date. In contrast to proven reserves, the total of 291 billion barrels is divided almost equally between OPEC and non-OPEC. The largest portion (46%) of past production has been in North America. Adding proven reserves and cumulative production, some 846 billion barrels of oil had been discovered in WOCA by 1975, and about one-third had been consumed.

Additions to Proven Oil Reserves 1950-1975

Before considering the rate at which proven reserves might be increased over the next 25 years, it is necessary to look at what has happened in the past 25 years. Two methods of considering past additions to reserves are advocated: one backdates estimates of reserves to the year in which they were discovered; the other compares year-end proven reserves figures as published in consecutive annual surveys of proven reserves in *Oil and Gas Journal* or *World Oil.* There are advantages and disadvantages in each method.

The backdating method attributes the estimated proven reserves in an oil field to the year it was discovered. For example, a field discovered in 1960 might be originally estimated to contain 1 billion barrels of recoverable reserves. Five years later, information obtained in developing the field might result in a re-estimate of the recoverable reserves as 1.2 billion barrels. The backdating method would attribute the extra 0.2 billion barrels to discoveries in 1960. Comparing year-end estimates of proven reserves would attribute this extra 0.2 billion barrels to additions to reserves in 1965.

Figure 3-2 is the picture that the backdating procedure gives for additions to reserves between 1930 and 1975. It also shows how these additions are divided between the Middle East and in areas out-

Figure 3-2 Oil Discovery Rate In WOCA from 1930 (Five-Year Average Obtained by Backdating Discoveries to Year of Field Discovery)

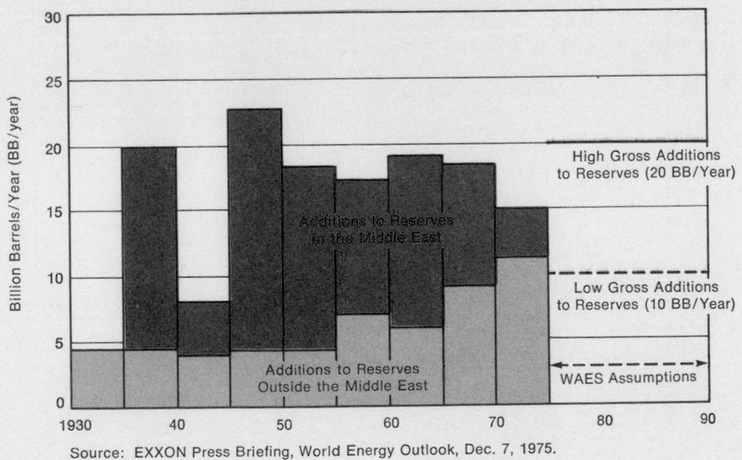

Source: EXXON Press Briefing, World Energy Outlook, Dec. 7, 1975.

side the Middle East. The rate at which reserves were discovered outside the Middle East was relatively constant—around 5 billion barrels per year—between 1930 and 1955, with most discoveries made in the U.S.A., Venezuela and Canada. The rate increased slowly to around 10 billion barrels per year by 1975, as reserves were found in areas such as Libya, Algeria, Nigeria, Indonesia, and, in the past decade, in Alaska and the North Sea.

These discoveries are overshadowed by those in the Middle East, which now contains around 60% of proven oil reserves in WOCA. Between 1950-1970, the average discovery rate in WOCA was about 18 billion barrels per year. Between 1970 and 1975, this rate fell to 15 billion barrels per year. This may be revised as further information on these recent discoveries becomes available.

Figure 3-3 shows a similar picture prepared by comparing annual estimates of proven reserves as published by *Oil and Gas Journal*. As in Figure 3-2, five-year averages are used to smooth out large annual fluctuations which can occur. The figure covers the period 1950-1975, since data are not available before that. It shows a different picture from that obtained by backdating. Additions to reserves averaged about 22 billion barrels per year between 1950

Figure 3-3 Oil Discovery Rate in WOCA (from 1950-1975) (Five-Year Average Obtained by Comparing Year-End Estimates of Proven Reserves as Published in Oil and Gas Journal)

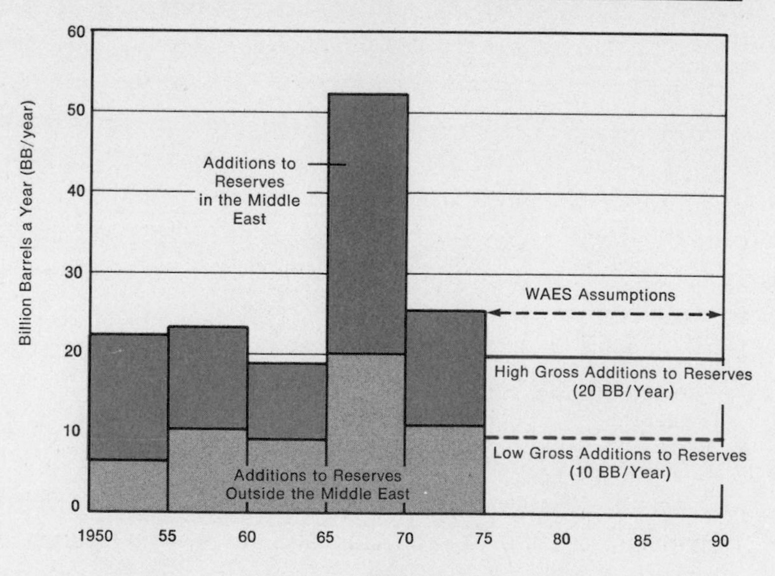

and 1965, but they increased to over 50 billion barrels per year in the five years between 1965 and 1970. Though the fields had actually been discovered before 1950, this was the period when the magnitude of the Middle East reserves, especially in Saudi Arabia, was being realized. Since 1970, additions to reserves have been much lower—around 25 billion barrels per year. Overall, additions to reserves averaged about 27 billion barrels per year between 1950 and 1975.

In 1950, the *Oil and Gas Journal* estimated proven reserves of WOCA at 72 billion barrels. Yet, as Figure 3-2 indicates, we now know that the reserves in oil fields discovered before 1950 were about 300 billion barrels. Thus, about 230 billion barrels of oil, which backdating attributes to pre-1950 discoveries, are included in the additions to reserves between 1950 and 1975 if comparison of annual proven reserve estimates is used.

The increase is not surprising. Oil exploration and production techniques improved rapidly in the period. New techniques increased the average rate of recovery from 25% to 32% in the U.S.A., for example. Seismic techniques made tremendous progress in the 1960's, resulting in a much earlier and more accurate estimate of recoverable oil in a new field. Also in the 1950's, there was a surplus in oil-production capacity. In that situation, there was little incentive for oil explorers to speed the evaluation of new reserves, especially in the Middle East, where oil production was already undergoing rapid development and fully able to meet rising demand.

Which of these two methods is best? Warman (1) states: "In my opinion the most meaningful analysis is obtained by taking the presently known reserves and backdating them to the year of discovery. It is my contention that nowadays the size of reserves is known fairly accurately to the operating company within a short time of the discovery of a field, and the size of the reserves fairly rapidly becomes generally known." Yet, even in the estimates of addition to reserves between 1970 and 1975, we see that new discoveries would be estimated at 15 billion barrels per year by the backdating method and 25 billion barrels per year by comparing annual year-end reserve estimates.

Therefore, in looking at future discovery rates we should look at results from both these methods. Backdating suggests that the rate at which new discoveries have been made has remained remarkably constant since 1950. Comparing annual proven reserves estimates suggests there were important additions as reserves in fields discovered before 1950 were reassessed.

In making estimates for the period after 1975 we should consider gross additions to reserves, by which we mean the total of:

(i) Reserves in genuine new discoveries

(ii) Additions to reserves in fields discovered before 1975, either through reassessment of reserves or enhanced recovery of the oil in place.

Future Gross Additions to Reserves

Genuine New Discoveries

The rate at which genuine new discoveries were made in the past would have been much lower without the discovery of the massive oil reserves of the Middle East. In an area some 800 miles by 500 miles, about 60% of WOCA oil reserves have been found. Is it likely that such a prolific oil-bearing region will be found again? The chances appear small. Many of the remaining possible areas have been evaluated by seismic techniques or exploration wells. No evidence of a new Middle East has been found.

It is likely that further new oil discoveries will be made in the Middle East, as Figure 3-2 shows, but there is already evidence that the rate at which new reserves are being found is beginning to decline. This is in spite of continuing drilling activity in the area. With only a small chance of either discovering a new Middle East or discoveries in the Middle East as large as in the past, the past rate of genuine new discoveries can only be achieved if a large number of smaller producing areas are found.

There are major uncertainties as to where new producing areas might be found. These uncertainties are well-illustrated by two examples. In the early 1960's, few people would have forecast the extent of the North Sea reserves, which by 1975 had proven reserves of about 23 billion barrels. Conversely, after Prudhoe Bay, Arctic Canada was forecast as a possible major oil province. Yet, despite a great deal of seismic work and numerous exploration wells, no commercial oil has been found.

As discoveries of oil are made, the reserves remaining to be found must decline since we are dealing with a finite resource base.

New techniques of assessing prospective regions make it possible to identify structures that might contain large oil accumulations. These are the areas that will be drilled first, and, as the search moves into less likely areas, the discovery rate will probably decline. The importance of finding large fields is shown by Table 3-3. Over 30,000 oil fields have been discovered, but about 75% of the oil lies within 240 large fields, each with over 500 million barrels of recoverable reserves. All the effort put into oil exploration around the world over the past one hundred years has only yielded 240 large oil fields. Yet, in hostile environments such as the North Sea, 500 million barrels of recoverable reserves is probably about the minimum size to justify the very costly production platforms and undersea pipelines needed to bring the oil ashore.

It therefore seems unlikely that genuine new discoveries will maintain even the 15 billion barrels-per-year rate achieved over the past five years.

Table 3-3 Approximate Size of World Oil Fields

	Number of fields discovered	*Estimated % of WOCA reserves*
A. All fields	30,000	100
B. Fields greater than 0.5 billion bbl. recoverable reserves	240	73
C. Fields greater than 10.0 billion bbl. recoverable reserves	15	34
D. 4 largest fields: Ghawar (Saudi Arabia), Greater Burgan (Kuwait), Bolivar Coastal (Venezuela), Safaniya-Khafji (Saudi Arabia/Neutral Zone)	4	21

Additions to Reserves in Fields Discovered Before 1975

Estimating recoverable oil reserves is much more accurate today than in the 1950's, and the corrections of the past 25 years resulting in large additions to known fields are unlikely to be repeated on the same scale.

A much more likely source of future additions to reserves is improvement of recovery techniques. The percentage of oil in place that can be recovered varies from field to field. Where the oil is heavy (viscous) recovery rates may be under 10% (with over 90% left in the ground). Conversely, in some rare cases where tertiary recovery is very successful, recovery rates have reached over 80%. It is extremely difficult to estimate recovery factors on a global basis, since published information relates to recoverable reserves without stating recovery factors. However, in the U.S.A. recovery currently averages 32%, and the figure for WOCA is probably between 30 and 35%.

It is uncertain how the recovery rate might increase in the future. Secondary recovery techniques have been used for several years, and in many cases oil from secondary recovery is now included when estimates of recoverable reserves are made. Tertiary recovery is a newer technique, which has been used with varying success in the U.S.A. Overall, the effect to date has probably been small. The increase in oil prices in the winter of 1973-74 certainly increased the incentive to use secondary and tertiary techniques, which may make an increasing contribution in the future.

If we make a conservative assumption that global recovery rates are currently 30%, then each percentage point increase in the recovery factor would increase proven reserves by a thirtieth. Proven reserves were 555 billion barrels at the end of 1975, while cumulative production was 291 billion barrels. Applying such an increase to all the proven reserves plus half the cumulative production would give additions to reserves of about 25 billion barrels from a percentage point increase in the recovery factor. If achieved before 2000, this would be equivalent to an annual rate of addition to reserves of 1 billion barrels per year.

Moody & Geiger (2) state that their figure of ultimately recoverable reserves of 1,500 billion barrels in WOCA assumes that global recovery factors will reach 40%. Such an improvement might eventually add 250 billion barrels to reserves in known fields, and if achieved before 2000, would give average additions to reserves of 10 billion barrels per year between 1975 and 2000.

However, it seems unlikely that such an increase could occur before 2000 without a major breakthrough in technology. It seems more likely that recovery rates will improve gradually and that annual additions to reserves from enhanced recovery may initially be small but increase over the next 25 years.

WAES Assumptions for Gross Additions to Reserves

Based on the preceding analysis it was decided in the WAES projections to use two future rates of gross additions to reserves: 20 billion barrels per year and 10 billion barrels per year.* The higher rate assumes that future exploration will be relatively successful, but that the discovery rate will gradually decline from the average of 18 billion barrels per year as the reserves that are yet to be discovered are reduced. To offset this effect, the contribution from enhanced recovery is projected to increase, giving the overall average rate of additions of 20 billion barrels per year. It is impossible to give exact figures, but between 1975 and 1985, possibly 75% of the additions will come from genuine new discoveries and 25% from known reserves. By the end of the century the proportions could perhaps be 50% from new discoveries and 50% from enhanced recovery from known reserves.

The lower rate of 10 billion barrels per year gross additions to reserves assumes that the rate of new discoveries will decrease rapidly and that enhanced recovery from fields discovered before 1975 will be relatively small. The division between genuine new additions and from enhanced recovery from known reserves might be similar to that assumed for the higher rate.

We also looked at the effect of a case that we consider unlikely—average gross additions to reserves of 30 billion barrels per year, which is slightly above the rate observed by comparing historical annual estimates as given by the *Oil and Gas Journal*. This case is discussed at the end of the chapter.

These average rates of additions to reserves are assumed for the period 1975-2000. Looking beyond 2000, we have to consider how they might decline as the figure for total ultimately recoverable reserves is approached. The determination of the rate of additions to reserves after 2000 is highly speculative, but Figure 3-4 is one attempt to show what might happen. It shows that cumulative discoveries

---

* It was decided to use these rates of gross additions to reserves in April 1975 as averages for the 25-year period 1975-2000. Since then new estimates of proven reserves have been made by *Oil and Gas Journal* which indicate gross reductions in 1976 of 40 billion barrels. Thus, to achieve the average of 20 and 10 billion barrels for the period 1975-2000 will, in fact, require additions to reserves of about 23 and 12 billion barrels per year over the 24-year period 1976-2000.

Figure 3-4 Cumulative Oil Discoveries in WOCA

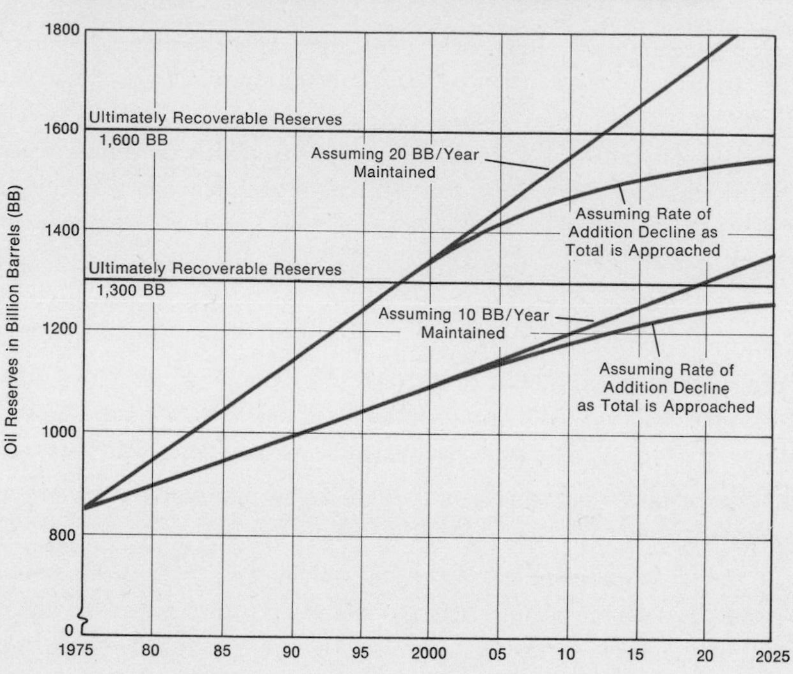

would exceed 1,600 billion barrels from conventional sources by 2013 if additions to reserves continued at 20 billion barrels per year after 2000. Thus the actual rate will probably decline after 2000, so that cumulative discoveries slowly tend towards ultimately recoverable reserves.

The lower rate for gross additions to reserves—10 billion barrels per year—could be maintained until 2050 before cumulative discoveries passed 1,600 billion barrels. However, this lower rate should perhaps be associated with a lower ultimate recoverable reserve of, say 1,300 billion barrels in WOCA (or 1,600 billion barrels in the total World). In this case the annual rate of gross additions to reserves might start to decline around 2000 as shown in Figure 3-4.

Therefore, we assumed that annual gross additions to reserves will follow the curves shown in Figure 3-4. The average annual rates in billion barrels per year are as follows:

	High gross additions to reserve case	*Low gross additions to reserve case*
1975-2000	20	10
2000-2010	12	8
2010-2020	7	6
2020-2025	4	3

The WAES study is for the period 1975-2000. In the case of oil we have looked beyond 2000 to obtain a picture of the oil profile over a 50-year period.

Oil Demand

The supply-demand integration described in Chapter 8 gives an estimate of *unconstrained* oil demand, i.e., oil demand assuming oil supplies are available. In this study of oil supply, it is assumed that WOCA unconstrained oil demand can be met by production for as long as is theoretically possible (limited only by technical factors such as the R/P ratio). When oil production can no longer increase, then oil demand will be constrained to the level of production.

WOCA oil demand as projected by the unconstrained supply-demand integration is as follows:

	Million Barrels per Day			*Average annual percentage growth rate*	
Scenario	*1975*	*1985*	*2000*	*75/85*	*85/2000*
C-1	45	63	93	3.4	2.6
C-2	45	63	92	3.4	2.5
D-7	45	58	76	2.5	1.8
D-8	45	58	75	2.5	1.8

Between 1960 and 1972, oil demand growth in WOCA averaged 6.2% per year. The projected growth rates in the WAES cases represent a significant decline from historical rates.

Because there is so little difference in the unconstrained oil demand in 2000 for Scenarios C-1 and C-2 and for Scenarios D-7 and D-8, it was decided to consider only C-1 and D-8. The results of Scenario C-1 will be used for Scenario C-2 and the results for Scenario D-8 for Scenario D-7.

Coupling of Oil Demand and Oil Discovery Rates

As described in Chapter 1, Scenario C-1 assumes high economic growth rates and oil prices that remain constant in real terms until 1985, gradually increasing 50% by 2000. Scenario D-8 assumes low economic growth and constant oil prices in real terms until 2000.

The assumptions about future gross additions to reserves depend to some extent on the size of the world oil resources. For example, additions to reserves cannot average 20 billion barrels per year over the next 25 years if ultimately recoverable reserves are only 1,300 billion barrels. This suggests that assumptions about additions to reserves may not be related to particular economic scenarios. However, in order to make a consistent study of the interaction between oil supply and demand, such a relationship has to be assumed.

In the case of high additions to reserves, we assume that enhanced recovery might add as much as 10 billion barrels a year by the end of the century. Enhanced recovery is more likely to occur in the rising-price than in a constant-price scenario. Therefore, the following couplings are adopted for this study: high gross additions to reserves (20 billion barrels per year) with the high-growth/rising-price scenario (C-1), and low rate of gross additions to reserves (10 billion barrels per year) with the low-growth/constant-price scenario (D-8).

The couplings are somewhat arbitrary, but they do represent two different views of oil and supply and demand. There are other possibilities. It is possible, for example, that the WOCA oil resource base is only 1,300 billion barrels, and that even with high economic growth and rising oil prices, exploration would be unsuccessful and yield low gross additions to reserves. On the other hand, a high gross addition to reserves might occur in a low-growth/constant-price world. This chapter will consider in detail the C-1/high-additions and the D-8/low-additions couplings but will also discuss possible effects of alternative couplings.

Table 3-4 summarizes these basic assumptions for WAES scenarios C-1 and D-8. The resulting oil supply curves are shown in Figure 3-5 (Scenario C-1) and Figure 3-6 (Scenario D-8).

(a) *Scenario C-1 (Figure 3-5)*

With the high gross additions to reserves, production ceases to meet demand in 1997 when the R/P ratio falls to 15 to 1. Thereafter production is limited to maintain a 15 to 1 R/P ratio and falls from the peak of 86 MBD in 1997 to 81 MBD in 2000 and around 32 MBD in 2025.

Figure 3-5 also shows the effect of a low gross additions to reserves for this scenario. Production first ceases to meet demand in 1990 when it peaks at 72 MBD. Production then declines to 50 MBD in 2000 and around 22 MBD in 2025.

Figure 3-5 WOCA Oil Supply—No Government-Imposed Limits on Production (Scenario C-1)

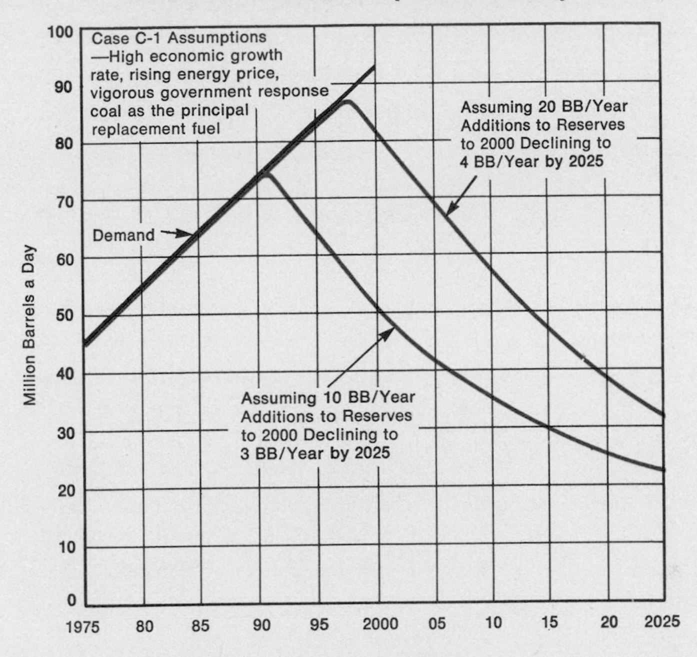

Figure 3-6 WOCA Oil Supply—No Government-Imposed Limits on Production (Scenario D-8)

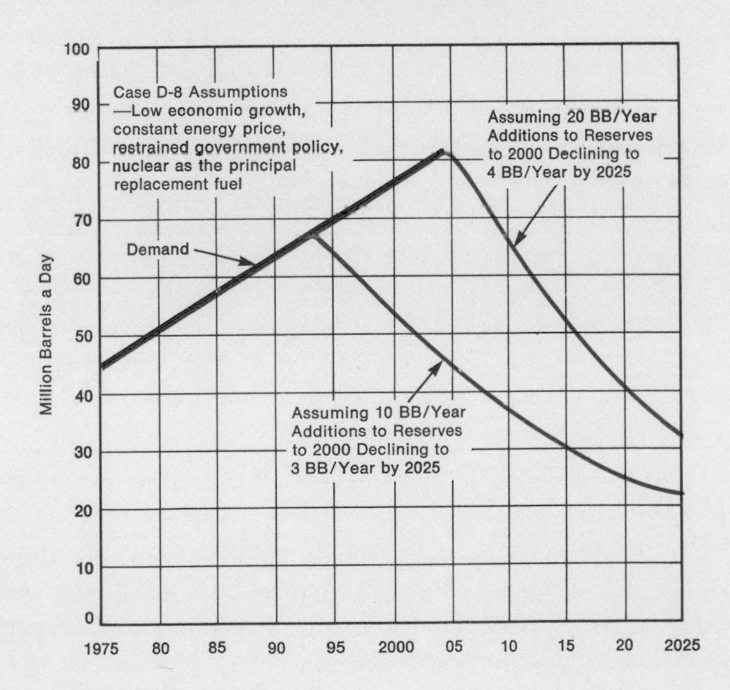

(b) *Scenario D-8 (Figure 3-6)*

With the low gross additions to reserves, oil production first ceases to meet demand in 1994. Thereafter production is limited to maintain a 15 to 1 R/P ratio and falls from the peak of 66 MBD in 1994 to 53 MBD in 2000 and just over 20 MBD in 2025.

Table 3-4 Assumptions for Oil Supply Profile

Scenario C-1

1.	Proven reserves in WOCA at 31st December 1975		— 555 billion barrels
2.	WOCA oil demand in 1975		— 45 MBD (16.5 billion barrels per year)
3.	WOCA oil demand growth	1975-1985	— 3.4% per year
		1985-2000	— 2.6% per year
4.	Gross additions to reserves	1975-2000	— 20 billion barrels per year
		2000-2025	— declining to 4 billion barrels per year by 2025
5.	Limiting R/P ratio		— 15 to 1

Scenario D-8

1.	Proven reserves in WOCA at 31st December 1975		— 555 billion barrels
2.	WOCA oil demand in 1975		— 45 MBD (16.5 billion barrels per year)
3.	WOCA oil demand growth	1975-1985	— 2.5% per year
		1985-2000	— 1.8% per year
4.	Gross additions to reserves	1975-2000	— 10 billion barrels per year
		2000-2025	— declining to 3 billion barrels per year by 2025
5.	Limiting R/P ratio		— 15 to 1

Note: Oil demand growth assumptions compare with the following for 1960-1975

1960-1972	6.2% per year
1972-1975	0.5% per year

Figure 3-6 also shows the effect of a high gross addition to reserves in this scenario. With such high additions, WOCA desired demand for oil of 75 MBD in 2000 can be satisfied. If demand were to continue to increase at 1.8% per year after 2000, then production could meet demand until 2004. Supply would peak at 81 MBD and then decline to around 30 MBD by 2025.

The Real World

In the foregoing pages, we have developed oil supply curves for WOCA that are based on assumptions about proven reserves, rate of additions to reserves, oil demand, and R/P ratios. It has been assumed that all countries will allow production to increase to meet demand and that the only limit on production would be an R/P ratio of 15 to 1.

It seems very unlikely that WOCA production would actually develop in this way. Some nations will want to reduce dependence on imported oil and will therefore produce domestic oil at a faster rate than assumed under a 15 to 1 R/P ratio. For example, we see that in 1975 the U.S.A. was already moving to a ratio of around 10 to 1.

For strategic and conservation reasons, other nations will restrict oil production well before a 15 to 1 R/P ratio is reached. In Iran, for example, a limiting R/P ratio of 25 to 1 is considered more likely, while some Saudi Arabian spokesmen have suggested that in certain circumstances production might be limited to 8.5 MBD. The latter limit is in spite of proven reserves in Saudi Arabia that could probably support a production rate of over 20 MBD.

Thus it is evident that the supply profiles in Figures 3-5 and 3-6 represent a most optimistic production profile designed to meet potential oil demand for as long as possible. They give an unlikely picture of WOCA oil supply and have therefore been labled "WOCA Oil Supply—No Government-Imposed Limits on Production." To obtain a more likely oil supply profile we must attempt to disaggregate the WOCA picture and look at production by region.

OPEC and Non-OPEC

A first step in disaggregating oil producers is to look at OPEC and non-OPEC countries separately. Most oil exporting countries belong to OPEC while most of the countries outside OPEC are oil importers.

Although all 13 members of OPEC are developing countries, their attitudes and dependence on oil production differ greatly. Nigeria, Iran, Indonesia and Venezuela, for instance, have relatively large populations, and need high oil revenues to support economic development programs. These countries must balance the need for immediate income against the desire to prolong the flow of oil revenues by limiting production. On the other hand, Saudia Arabia, the United Arab Emirates, Kuwait and Libya are currently earning oil revenues well in excess of domestic requirements, and therefore have little incentive to increase oil production above present levels.

The "non-OPEC" group contains countries at all stages of economic development and with various degrees of dependence on imported oil. With the possible exception of a few countries such as the U.K., Norway and Mexico, they are likely to remain oil importers in the future. Consequently, the majority of them can be expected to maximize domestic oil production in order to reduce their dependence on OPEC oil.

Because of these important differences, it is necessary to consider OPEC and non-OPEC aggregations separately, and where possible, consider individual countries in detail. But we should first discuss how gross additions to reserves might be divided between OPEC and non-OPEC in the future. Table 3-2 (page 118) shows that total past discoveries in WOCA have been as follows (in billion barrels):

	Non-OPEC	*OPEC*
Cumulative production to date	152	139
Remaining proven reserves	105	450
Total	257	589

Although production to date has come almost equally from OPEC and non-OPEC, about 70% of WOCA discoveries have been in OPEC, which now contains over 80% of remaining reserves.

Warman (1) and Moody & Geiger (2) give estimates of the potential for future discoveries by region but not divided between OPEC and non-OPEC. We have assumed that reserves still to be discovered might be equally divided between OPEC and non-OPEC. This should be seen as a working assumption. Within the non-OPEC area, there are countries where there has been little exploration and where many new oil fields may be found. However, the gross additions to reserves include enhanced recovery and additions to known reserves revealed by further development. These will be largest in OPEC where most of the currently proven reserves are located.

Table 3-5 Estimates of Non-OPEC WOCA Oil Production 1975-2000* (in MBD)

		Case C	*Case C-1*
	1975	*1985*	*2000*
North American	11.7	14.0	9.3
Western Europe	0.6	4.5	2.9
Japan	0.01	0.03	0.4
Non-OPEC Rest of WOCA	4.0	6.2	12.0
Total Non-OPEC	16.3	24.7	24.6

		Case D	*Case D-8*
	1975	*1985*	*2000*
North America	11.7	11.9	8.0
Western Europe	0.6	4.5	2.6
Japan	0.01	0.03	0.4
Non-OPEC Rest of WOCA	4.0	5.6	7.5
Total Non-OPEC	16.3	22.0	18.5

* Production estimates for North America, Western Europe and Japan from WAES national teams.

Non-OPEC Production 1975-2000

Each national team in WAES has produced an oil production estimate for their own country (see WAES Technical Report No. 2).* For the non-WAES developing countries, oil production estimates are summarized in Appendix 1 to this report. The estimates for the non-OPEC, non-Communist world by region are shown in Table 3-5.

The 1985 figures are determined largely by reserves already known and under development. The long lead times involved in exploring and developing new oil regions in difficult environments, such as the North Sea and Alaska, mean that discoveries made between now and 1985 will probably make only a marginal contribution to production in 1985. This is especially true in Western Europe, where most of the production will be from offshore in the North Sea. Production in Western Europe will therefore be largely unaffected by economic growth or oil price, and similar production levels are assumed for both Scenario C-1 and Scenario D-8. In North America and in the developing world, there is hope that a more favorable economic climate could lead to increased development. This is reflected

* *Energy Supply to the Year 2000: Global and National Studies.*

in the higher production forecasts for Scenario C-1, compared with Scenario D-8. Japanese production is expected to remain of marginal importance in both scenarios.

The situation is much more uncertain for 2000. Oil reserves that are being produced now will, to a large extent, be exhausted and an increasing share of production will come from discoveries made after 1975. The rate of discovery becomes increasingly important and results in different production rates for each scenario, especially in North America and the developing world. In Europe, reserves in the North Sea—the one major oil region—are limited, and no further major discoveries are expected. Thus, the Western Europe production forecasts are similiar for each scenario.

Non-OPEC Production 2000-2025

The WAES national studies in general have not looked beyond 2000. It is therefore necessary to resort to the methods used earlier in this chapter to consider the potential production rates from proven reserves given limiting R/P ratios.

The remaining non-OPEC reserves in 2000 depend on the oil production and the gross additions to reserves between 1975 and 2000. In Scenario C-1, where the additions are assumed to be high, with half occurring in the non-OPEC region, the remaining non-OPEC reserves would be about 130 billion barrels in 2000 and the R/P ratio around 14 to 1. The comparable figures for Scenario D-8 are 76 billion barrels and a reserves to production ratio of 11 to 1.

After 2000, the rate of addition to reserves is likely to be in decline since the undiscovered reserves are reduced. The R/P ratio may also decline as demand increases pressure on the remaining reserves. A production forecast based on declining gross additions to reserves and a declining R/P ratio might look like this:

Non-OPEC Potential Production Rates: 2000-2025
(in MBD)

	Scenario C-1	*Scenario D-8*
2000	25	19
2005	23	17
2010	21	15
2015	19	13
2020	16	11
2025	13	9

OPEC Production 1975-2000

To achieve the WOCA oil production profile shown in Figures 3-5 and 3-6 would require the following production rates from OPEC.

Required OPEC Production Rates (in MBD)

	Scenario C-1 (High additions to reserves High Demand)	*Scenario D-8* (Low additions to reserves Low Demand)
1975 (actual)	27	27
1980	32	29
1985	39	36
1990	47	42
1995	57	44
2000	56	34

These rates assume that all OPEC countries will allow production to increase to meet WOCA oil demand as long as the R/P ratio is at least 15 to 1. The rates required, especially in the C-1 case, are well in excess of current capacity of around 37 MBD. It is doubtful whether if they are really feasible either on technical or policy grounds. Iran, for example, may feel a 25 to 1 R/P ratio might be more realistic, while the expansion in oil production infrastructure in some other countries is unlikely to take place on the scale required.

Individual OPEC countries may also be unwilling to see oil production reach the maximum theoretical level because of a desire to extend the life of their oil reserves. Already some OPEC governments have expressed such views by announcing possible limits to oil production as shown in Table 3-6.

Table 3-6 Possible Production Limits Announced by Governments

Country	*Estimated Usable** *Capacity at End 1975* (in MBD)	*Government-Announced Possible Production Limits* (in MBD)
Venezuela	2.5	2.2
Ecuador	0.22	0.2
Libya	2.5	2.0
Qatar	0.7	0.5
United Arab Emirates	2.3	1.8
Kuwait	3.0	2.0
Saudi Arabia	10.8	8.5
TOTAL	22.0	17.2

* As given in *Petroleum Intelligence Weekly*, 2 February 1976.

If imposed, these government restrictions on production would reduce OPEC-usable capacity by some 4.8 MBD from the estimated usable capacity of 37 MBD in 1975. The majority of these restrictions are in countries that are "low absorbers" of oil revenues: Kuwait, Saudi Arabia, Abu Dhabi, Qatar and Libya. Venezuela and Ecuador have imposed restrictions to conserve declining reserves and prolong the flow of oil revenues, even though these countries could probably absorb all the oil revenues they could earn.

The real impact of government-imposed restrictions, however, is on future potential production rates. Table 3-7 summarizes the present oil reserves, oil production and usable capacity in the OPEC countries. It shows that the OPEC countries of the Arabian Peninsula (Saudi Arabia, Kuwait, Qatar and the United Arab Emirates) dominate

Table 3-7 OPEC Oil Production Statistics

	Reserves at[1] *1.1.76*	*Production*[2] *1975*		*R/P Ratio*	*Estimated Usable Capacity*[2]
	(billion barrels)	(MBD)	(in billion barrels per year)		(MBD)
Middle East					
Saudi Arabia	152.1	7.1	2.59	59	10.8
Kuwait	71.3	2.1	0.77	93	3.0
United Arab Emirates	32.2	1.7	0.62	52	2.3
Qatar	5.9	0.4	0.15	40	0.7
Iran	64.5	5.4	1.97	33	6.8
Iraq	34.3	2.3	0.84	41	2.6
Total Middle East	360.3	19.0	6.94	52	26.2
Africa					
Nigeria	20.2	1.8	0.66	31	2.7
Libya	26.1	1.5	0.55	48	2.5
Algeria	7.4	1.0	0.37	20	1.0
Gabon	2.2	0.2	0.07	30	0.2
Others					
Venezuela	17.7	2.3	0.84	21	2.5
Indonesia	14.0	1.3	0.47	30	1.7
Ecuador	2.5	0.2	0.07	34	0.2
Total Africa and Others	90.1	8.3	3.03	30	10.8
Total OPEC	450.4	27.3	9.97	45	37.0

[1] *Oil and Gas Journal*, 29 December 1975.
[2] *Petroleum Intelligence Weekly*, 2 February 1976.

OPEC reserves with 261 billion barrels, or 58% of total OPEC-proven reserves. The R/P ratios of these Arabian Peninsula countries are, with the exception of Qatar, in excess of 50 to 1. If a ratio of 15 to 1 were the only limit on production, they could today produce oil at a rate of 47 MBD (equivalent to current total WOCA output). This compares with a possible government-imposed production list in 1976 of 12.8 MBD, or some 4.0 MBD below current installed capacity.

The remaining countries with government-imposed production limits have a much smaller potential to expand production. Currently the restrictions reduce usable capacity by 0.8 MBD. Such limits will have little effect on the conclusions of a long-term study such as ours, in spite of the possible implications for the countries concerned. It is the effects of any government limits imposed by countries with major potential for expanding oil production—Saudi Arabia, Kuwait and the United Arab Emirates—that will be important and which need to be considered in more detail.

The remaining OPEC countries outside the Arabian Peninsula have estimated usable capacity of 20.2 MBD, but possible government limits as shown in Table 3-6 could reduce this by 0.8 million to 19.4 MBD. The reserves of some of these countries are already being reduced fairly rapidly (the R/P ratios in Algeria and Venezuela are already down to 20 to 1, while Iran is nearing its probable limit of 25 to 1). New additions to reserves in these countries are likely to enable them to maintain production at present levels rather than to significantly increase it. We therefore assumed a limit of 20 MBD on the oil production from OPEC countries outside the Arabian Peninsula. This limit, which is similar to present capacity, takes into account both the physical limits imposed by reserves and any other limits that may be imposed by governments. It is an assumption rather than a forecast, and will be used in both the high and low additions to reserves cases.

For the Arabian Peninsula countries (Saudi Arabia, Kuwait, Qatar, and the United Arab Emirates) three assumptions were made:

(1) Government limits at near present production levels. This would limit output to around 13 MBD.

(2) Saudi Arabia allows production to increase to about 15 MBD, and as a result, total Arabian Peninsula potential production increases to 20 MBD.

(3) Saudi Arabia allows production to increase to about 20 MBD, and as a result, total Arabian Peninsula potential production increases to 25 MBD.

Assumption (1) above will be considered in both the high and low additions to reserves cases. Assumption (3) would require a rapid increase in production capacity to lead the governments to adopt such an increase over current production levels and will be considered only in the high gross additions to reserves case. Assumption (2) probably requires a smaller increase in infrastructure and in reserves and will be considered only in the low gross additions to reserves case.

We therefore have three assumptions for limits to total OPEC production:

(i) 33 million barrels per day;
(ii) 40 million barrels per day; and
(iii) 45 million barrels per day.

These levels define the maximum potential production assumed to be available from OPEC. Once the required production from OPEC reaches such a limit, it will be assumed to remain at that level as long as the R/P ratio exceeds 15 to 1. From then on, maintaining a 15 to 1 R/P ratio will result in declining OPEC production.

The resulting OPEC production rates from 1975-2000 are shown in Table 3-8.

Table 3-8 Projected OPEC Production Rates 1975-2000 (in MBD)

	Scenario C-1 High Gross Additions to Reserves (10 billion barrels/year in OPEC)			*Scenario D-8* Low Gross Additions to Reserves (5 billion barrels/year in OPEC)		
	OPEC Limit 33 MBD	OPEC Limit 45 MBD	OPEC no govt. limit	OPEC Limit 33 MBD	OPEC Limit 40 MBD	OPEC no govt. limit
1975	27	27	27	27	27	27
1980	32	32	32	29	29	29
1985	33	39	39	33	36	36
1990	33	45	47	33	40	42
1995	33	45	57	33	40	44
2000	33	45	56	33	39	34

In Scenario C-1, the 45 MBD rate of production is reached in 1989 and the lower limit of 33 MBD in 1981. In Scenario D-8 the 40 MBD limit is also reached in 1989 and the lower limit in 1983.

OPEC Production 2000-2025

Beyond 2000, the OPEC production rates are more speculative. They depend on the additions to reserves made between 1975 and 2000, the production rates that the countries outside the Arabian Peninsula can maintain, and the attitudes of Arabian Peninsula countries toward restraints. If we assume that OPEC limits of 33, 40 and 45 MBD are maintained—if necessary by some countries relaxing restraints—then the production rates shown in Table 3-9 might be possible.

Table 3-9 OPEC Production Rates 2000-2025 (in MBD)

	Scenario C-1 High Additions to Reserves (10 billion barrels/year in OPEC)			*Scenario D-8* Low Additions to Reserves (5 billion barrels/year in OPEC)		
	OPEC Limit 33 MBD	OPEC Limit 45 MBD	OPEC no govt. limit	OPEC Limit 33 MBD	OPEC Limit 40 MBD	OPEC no govt. limit
2000	33	45	56	33	39	34
2005	33	45	45	33	31	26
2010	33	42	37	33	25	22
2015	33	33	28	32	20	19
2020	33	28	23	25	17	17
2025	33	22	19	20	13	13

Total World Outside Communist Areas' Oil Production

Combining the production estimates for non-OPEC countries and the assumed OPEC production levels gives estimates for total WOCA oil production.

Figure 3-7 shows WOCA oil production for Scenario C-1 with gross additions to reserves of 20 billion barrels per year. The profiles

Figure 3-7 WOCA Oil Production for C-1

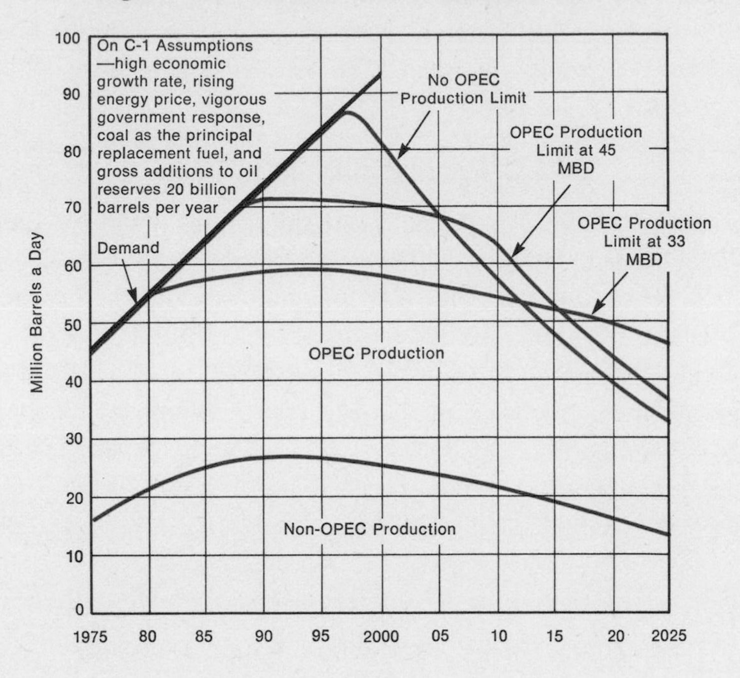

Figure 3-8 WOCA Oil Production for D-8

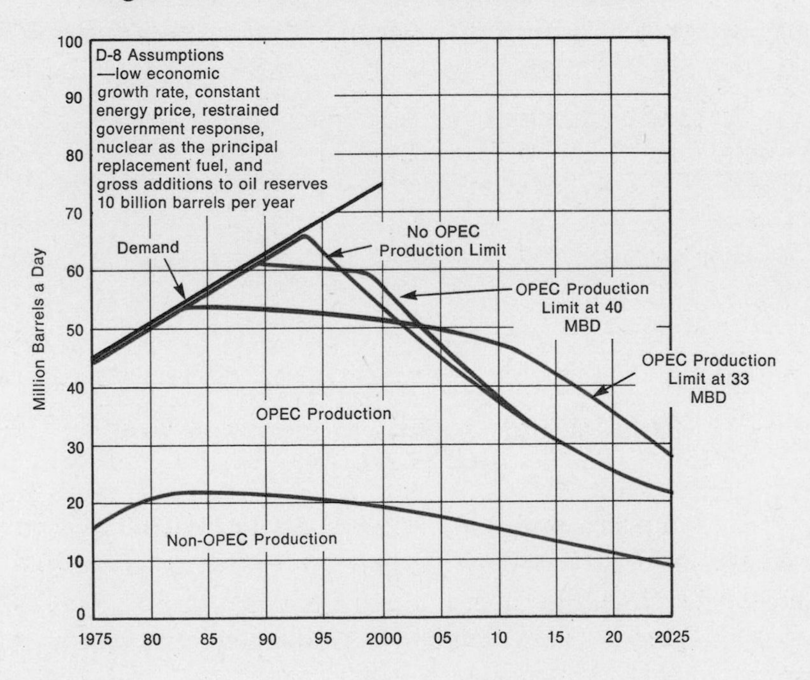

that assume maximum OPEC production of 33 and 45 MBD are contrasted with the profile assuming no government limits on production. The production profile for non-OPEC is also shown.

An OPEC limit of 45 MBD results in WOCA production being constrained by 1989, compared to 1997 for the unlimited case. OPEC production can be maintained at 45 MBD until 2007. For nearly 20 years, the total WOCA profile is nearly constant, with peak production of about 71 MBD or some 15 MBD below the unconstrained peak.

If OPEC maximum production were limited to 33 MBD, then WOCA oil production would be constrained as early as 1981. The limit of 33 million barrels could be maintained until 2025 and the only variations in WOCA production after 1981 would arise from changes in non-OPEC production. The result is a slow increase in WOCA production to a maximum of about 59 MBD in 1995 followed by a slow decline to 46 MBD by 2025.

Figure 3-7 shows clearly the impact of OPEC constraints. Oil is kept in the ground in the 1990's, increasing the remaining reserves. This oil can then be produced after 2000, making the decline in oil production less severe. An OPEC production limit of 45 MBD results in production in 1997 being 15 MBD lower than the maximum possible. However, production in the early part of the next century would be higher than if reserves are depleted at the fastest rate possible in the 1990's.

Figure 3-8 shows comparable WOCA oil supply curves for Scenario D-8, assuming that gross additions to reserves will be 10 billion barrels per year. With an OPEC production limit of 40 MBD, oil supply reaches a peak of 61 MBD in 1989, or some 5 million less than the unconstrained peak which occurs in 1994. OPEC can hold production at this level until 1999. The profile for WOCA between 1989 and 1999 declines slowly because of the decline in non-OPEC production. The impact of keeping the OPEC oil in the ground is already felt by 2000 since the reserves conserved enable production in 2000 to be 58 MBD or about 5 MBD above the level possible without OPEC constraints.

If OPEC production were to be limited to 33 MBD, then WOCA oil supply would be constrained as early as 1983. OPEC production could be held at this limit until 2013 so that between 1983 and 2013, total WOCA production declines slowly as non-OPEC production declines.

World Outside Communist Areas' Oil Supply/Demand Balance in 1985 and 2000

Potential oil demand in 1985, as projected in Scenarios C and D, can be met even if OPEC were to impose a limit on production of 40 or 45 MBD. However, if members of OPEC who have recently suggested possible production ceilings were to actually impose them, WOCA oil production could be constrained early in the 1980's. In this case there could be a gap between potential oil demand and oil supply in 1985 for both Scenarios C and D.

Table 3-10 summarizes the various cases considered through 2000. The only case in which potential oil demand can be met is if gross additions to reserves are high, oil demand is low and OPEC does not impose any limits on production. This is an unlikely set of events, and it is more probable that potential oil demand will not be met in 2000 on either Scenario C-1 or Scenario D-8.

In the high growth Scenario C-1, the shortfall in oil supply in 2000 is about 12 MBD if gross additions to reserves are high, and if OPEC countries impose no limits on production. OPEC limits of 45 MBD per day would increase this shortfall to about 23 MBD while more severe limits of 33 MBD would increase it to 35 MBD.

The effect of OPEC limitations of 40 MBD in Scenario D-8 with low gross additions to reserves in the opposite; they tend to reduce the projected shortfall in oil supply by 2000. Oil kept in the ground is, by 2000, already helping to defer the decline in production. Thus the shortfall of 22 MBD with OPEC production unconstrained could be lower—at 17 MBD—if OPEC production were not allowed to exceed 40 MBD. In the case of a lower OPEC limit of 33 MBD the shortfall in 2000 could be 23 MBD.

A More Optimistic Future?

The assumptions made in this chapter for gross additions to oil reserves may be considered by some observers to be conservative. Estimates that ultimately recoverable reserves may be 3,000 billion barrels or even higher have been published. If these estimates prove to be true, then gross addition to reserves may indeed be higher than our high case of 20 billion barrels per year.

More optimistic estimates tend to be based on the assumption

Table 3-10 Range of WOCA Peak Production

Scenario	*Gross Annual Addition to Reserves*	*OPEC Limits*	*Year production ceases to meet demand*	*Production in year when demand first ceases to be met*	*Production in 2000**	*Potential Demand in 2000*	*Possible Shortfall in Oil Supply*
	(in billion barrels)	(MBD)		(MBD)	(MBD)	(MBD)	(MBD)
C-1	20	None	1997	86	81	93	12
C-1	10	None	1990	72	50	93	43
C-1	20	33	1981	55	58	93	35
C-1	20	45	1989	71	70	93	23
D-8	10	None	1994	66	53	75	22
D-8	20	None	2004	81	75	75	0
D-8	10	33	1983	55	52	75	23
D-8	10	40	1989	61	58	75	17

* These oil production estimates do not include nonconventional oil sources such as oil sands and oil shale. In the year 2000 in the C-1 cases, these sources contribute 2.9 MBDOE and in the D-8 cases only negligible amounts. When these sources are included in the global supply-demand integrations, they thereby reduce the oil shortfall shown in the last column by the amount of their contribution.

that either genuine new discoveries will be high in the future or that enhanced recovery will make a major contribution in the future. To maintain our high gross additions to reserves assumption of 20 billion barrels will require a major successful exploration and development program. It would probably require new oil regions the size of the North Sea (proven reserves at the end of 1975 of 23 billion barrels) or Nigeria (20 billion barrels) to be discovered at much more frequent intervals than in the recent past. However, some estimates put the prospects for the developing countries much higher than we have. If oil exists in these areas and in the suggested volumes, then we might see a much higher level of genuine new discoveries than is assumed here.

In our high case, it is assumed that enhanced recovery may contribute up to 50% of the gross additions to reserves by 2000. To achieve such a figure would require major investment and success in research and development of secondary and tertiary recovery techniques. However, on average some 65%-70% of the oil in an oil field is at present left unrecovered. A major breakthrough in secondary and tertiary recovery techniques which substantially reduced this percentage, could make a contribution to additions to reserves even higher than we have assumed. Such a breakthrough would have to occur soon if a major contribution to oil supplies were to occur in this century. It therefore appears that improvements in enhanced recovery above those assumed in this chapter are more likely to extend the oil plateau into the 21st century than to substantially increase oil production before year 2000.

However, the possibility that gross additions to reserves may be higher than 20 billion barrels per year cannot be ruled out. We therefore considered the possible effects of additions to reserves averaging 30 billion barrels per year over the next 25 years, i.e., slightly higher than the figure of about 27 billion barrels per year between 1950 and 1975 obtained by comparing successive year-end estimates as published by the *Oil and Gas Journal.* If there were no constraints on OPEC production, such a rate of additions to reserves would allow production to meet potential demand in Scenario C-1 in 2000. Demand could continue rising for the first five years of the next century, but sometime before 2010, a limiting R/P ratio of 15 to 1 would be reached and production would peak and decline.

As has been shown in this chapter, the levels at which OPEC

countries are prepared to produce their oil is likely to be a much more important factor than the theoretical limits imposed by a 15 to 1 minimum R/P ratio. The year in which an OPEC limit of 45 million barrels is reached in this case depends on the level of production in non-OPEC countries. Even with a discovery rate outside OPEC of 15 billion barrels per year which might allow a rapid increase in non-OPEC production, OPEC production would still have to exceed 45 MBD soon after 1990 if oil production were to continue to meet demand. By 2000, with OPEC production at 45 MBD, production might be between 75 and 80 MBD. This is less than 10 MBD above production if additions to reserves were 20 billion barrels per year and still about 15 MBD below potential oil demand.

Conclusion

The future of oil supply is uncertain. The cases considered in this analysis show differing years and different levels of peak production, with dissimilar levels of production in year 2000. However, one conclusion is very clear: potential oil demand in the year 2000 is unlikely to be satisfied by crude oil production from conventional sources. Even if there are no governmental constraints on oil production, oil supply will meet demand only under the most optimistic assumptions about gross additions to reserves. A much more important constraint on oil production is likely to be the levels of production that are set by producers with large reserves and limited needs for more revenue. Possible constraints on oil production by members of OPEC are likely to cause oil supply to peak at the latest sometime around 1990 although lower production limits could bring this date forward into the early 1980's.

The end of the era of growth in oil production is probably at the most only 15 years away. However, there may be a decade or so of more or less constant oil production after 1990 in which consumers will have to make the adjustments necessary to face a decline in oil supply. It is possible that oil discoveries or recovery factors might be higher than we have assumed, but the effect would be to delay the impact for a few years rather than solve the problem of transition to other fuels. For nations to continue to increase their consumption of oil in the hope that more optimistic estimates might prove to be correct is to run the real risk that the peak in oil production could be brought forward, making the necessary adjustments in energy consumption patterns much more severe.

References:

(1) H.R. Warman, "Future Availability of Oil"—paper presented at Conference on World Energy Supplies 18th-20th September, 1973.

(2) J.D. Moody and R.E. Geiger, "Petroleum Resources: How Much Oil and Where?" *Technology Review*, March/April 1975.

CHAPTER 4

NATURAL GAS

Producing Gas — Proven Reserves — Ultimately Recoverable Reserves — Natural Gas Supply and Demand 1950-1972 — Natural Gas Supply and Demand 1972-2000 — Required Gas Imports in 1985 and 2000 — Potential Natural Gas Trade — Balancing Natural Gas Supply and Demand in 1985 — Balancing Supply and Demand in 2000

Natural gas is a clean and convenient fuel for cooking and heating and is a valuable raw material for the large and growing petrochemical industry. World reserves are large and unlikely to limit production within the next 25 years. However, the future role of gas as an energy source will be determined not so much by supply but by the problem of transporting and distributing gas from the wells to the consumer. Problems of transport have limited intercontinental trade in gas, which amounted to only .3 MBDOE in 1975.

Historically, natural gas has moved directly from supplier to consumer through pipelines. The large-scale use of gas has been in markets that could be economically connected by pipeline to natural gas reserves. The expense of constructing costly pipeline networks could only be justified where there are both large reserves and an assured demand. In contrast to oil, which supplies over a third of the energy in every major energy-consuming country, dependence on natural gas varies from virtually zero in Sweden, Denmark, and Japan to around 30% in the U.S.A. and 47% in The Netherlands.

An alternative to pipelines is transport by tanker as liquefied natural gas (LNG). The technology for liquefication, transportation and regasification has been available on a commercial scale only during the last decade. The gas is liquefied, then carried in special refrigerated tankers at −161°C, and regasified at receiving terminals. The method presents some problems. Approximately 25% of primary energy is lost in the processing of LNG. There is concern that an LNG tanker explosion in a port could cause severe damage to life and property. An LNG system also requires large capital investments for moving gas from the Middle East (where the largest unexploited reserves are) to Europe, Japan or the U.S.A. (the largest potential users). Because of the large investment in infrastructure for gas distribution and use in some countries, LNG may become increasingly desirable in order to avoid replacements based on other fuels.

In the following sections we describe how gas is produced and the size and distribution of the reserves. This is followed by our projections of future demand and supply of natural gas by region for the WAES scenario cases. The chapter concludes with a discussion of the potentials and problems of future international gas trade.

For consistency with the rest of the report the units used in this chapter are expressed in MBDOE. Natural gas is usually discussed in trillion (10^{12}) cubic feet (TCF), billion (10^{9}) cubic feet per day (BCFD) or in billion (10^{9}) cubic meters (BCM). The standard conversions from MBDOE into the other units are approximately:

1 MBDOE = 2.12 TCF/year;

" = 5.9 BCF/day; and

" = 57.0 BCM/year.

Producing Gas

Natural gas occurs in underground structures similar to those containing oil (see Chapter 3) and the methods for exploration and production of natural gas are similar. There are varying quantities of gas in most oil reservoirs, either in solution with the oil or in a gas cap above the oil. Such gas is called *associated gas.* When it is dissolved in the oil, gas is necessarily produced along with the oil. On the other hand, gas found in a cap above the oil is rarely produced until after the oil has been extracted. Earlier production of the gas would reduce the yield of oil. In some structures, only natural gas can be produced. This is called *unassociated gas*. Globally, about 40% of gas reserves are associated with oil, 60% are unassociated.

Proven Reserves

At the end of 1975, about 520 billion barrels oil equivalent of natural gas had been discovered—probably between 25 and 50% of the ultimately recoverable reserves. Of this total, 123 billion barrels oil equivalent had been consumed leaving about 400 billion barrels oil equivalent in proven reserves. Estimates of remaining proven reserves as of January 1976 are shown in Table 4-1 (from *Oil and Gas Journal*, 29 December 1975).

Two-thirds of the proven reserves are in WOCA and one-third is in the Communist world. Although the major share of WOCA's natural gas reserves are in OPEC, the distribution of gas reserves is much wider than oil reserves. In WOCA, 45% of the natural gas reserves lie outside OPEC countries (compared with only 20% for oil). About 30% of the reserves lie in the major energy-consuming areas of North America and Western Europe (compared with 12% for oil).

Table 4-1 Total World Proven Natural Gas Reserves

Region	*Remaining Reserves in 10^{12} Cu. Ft.*	*Remaining Reserves in Billion Barrels Oil Equivalent*
OPEC: Iran	330	57
Saudi Arabia	103	18
Other Middle East	96	17
Algeria	126	22
Other Non-Middle East	133	23
Total OPEC	788	137
North America	268	46
Western Europe	181	31
Rest of WOCA	160	28
Total Non-OPEC	609	105
Total WOCA	1397	242
Communist Areas	835	144
Total World	2232	386

Source: *Oil and Gas Journal*, Dec. 1975.

Ultimately Recoverable Reserves

Estimates of ultimately recoverable world natural gas reserves are subject to even more uncertainty than oil reserve estimates. This is partly because, in the past, much of the gas found was associated with oil and was treated as an unwanted by-product of oil production. Or it was found in areas too remote for it to be economically brought to market. Table 4-2 below, from Kirkby and Adams,[1] gives some estimates of ultimately recoverable natural gas reserves made over the last 20 years.

Table 4-2 Estimates of World Ultimately Recoverable Reserves of Natural Gas

Year Made	*Source*	*Reserves in 10^{12} Cu. Ft.*	*Reserves in Billion Barrels Oil Equivalent*
1956	U.S. Department of the Interior	5000	860
1958	Weeks	5000-6000	860-1035
1959	Weeks	6000	1035
1965	Weeks	7200	1240
1965	Hendricks (USGS)	15300	2640
1967	Ryman (ESSO)	12000	2070
1967	Shell	10200	1760
1968	Weeks	6900	1200
1969	Hubbert	8000-12000	1380-2070
1971	Weeks	7200	1240
1973	Coppack	7500	1300
1973	Hubbert	12000	2070
1973	Linden	10400	1800
1975	Kirkby and Adams	6000	1030
1975	Moody and Geiger	8150	1400

Source: Kirkby and Adams, Presentation at World Petroleum Congress, Tokyo, May 1975.

The estimates have been converted into billion barrels oil equivalent to facilitate comparison with estimates of oil reserves in Chapter 3. Estimates of ultimate recoverable reserves of oil increased until the mid-1960's. Since then, oil estimates have tended to converge around 2,000 billion barrels. No such convergence is seen in the estimates of gas reserves. Estimates made since 1965 have ranged from 1,030 to 2,640 billion barrels oil equivalent.

Estimates of ultimately recoverable reserves of gas and the rate at which these reserves are increased are less important than the location of the reserves. Currently proven reserves and possible future additions (even if ultimately recoverable reserves are as low as 1,030 billion barrels) are likely to allow for a substantial increase in natural gas production during the next few decades provided the gas can be moved to market.

Moody & Geiger[2] give their estimates of ultimately recoverable gas of 1,400 billion barrels oil equivalent by region. They show that only about 25% of the world's natural gas reserves are likely to be found near the main energy-consuming countries of WOCA. About 30% is expected to be found in the Communist areas, with the remaining 45% located in the Middle East and other developing countries. These estimates are shown in Table 4-3.

Table 4-3 World Ultimately Recoverable Natural Gas Reserves by Region

	Reserves in Billion Barrels Oil Equivalent	*% Share of Total*
North America	280	20.0
West Europe	77	5.5
Middle East	270	19.3
Rest of WOCA	348	24.9
Communist Area	425	30.3
Total World	1400	100.0

Source: "Petroleum Resources: How Much Oil and Where?" Moody and Geiger, *Technology Review*, March 1975.

Natural Gas Supply and Demand 1950-1972

An analysis by region of natural gas demand shows how the proximity of reserves has stimulated consumption. In North America, where large reserves of gas were discovered in association with oil and where prices have been low compared to oil, natural gas has become a major source of energy. In 1950, natural gas consumption in North America was 3.2 MBDOE, or 9% of total energy consumption. This was over 91% of world natural gas consumption in 1950. By 1975, gas consumption had grown to 11.3 MBDOE, providing a third of North American total primary energy and about 68% of the gas consumed in WOCA as shown in Table 4-4.

In the United States, over 40 million households and 3 million businesses rely on gas. Industry relies on gas both as an energy source and as a petrochemical feedstock. It has also been used for electrical generation in areas close to abundant gas reserves.

Table 4-4 WOCA Natural Gas Consumption in 1975

	MBDOE	*% of WOCA*
North America	11.3	68
Western Europe	3.1	19
Rest of WOCA	2.1	13
Total WOCA	16.5	100

Source: *BP Statistical Review*, 1975.

In Western Europe, the natural gas industry was of minor importance until the early 1960's when the discovery and development of the Groningen field stimulated gas consumption in The Netherlands and surrounding countries. Consumption was further stimulated by discoveries of gas in the southern North Sea in the mid-1960's, which provided the U.K. with a source of natural gas to replace gas manufactured from coal. Production from these gas fields allowed an increase in gas consumption in the U.K. from 2% of total energy in 1960 to 15% in 1975. Between 1969 and 1972, natural gas consumption in Western Europe increased at a rate of 30% per year when total energy was increasing only 5 or 6% per year. As Table 4-4 shows, natural gas consumption was 3.1 MBDOE in Western Europe in 1975—19% of the total natural gas consumption in WOCA.

There was almost no natural gas consumed in Japan until the late 1960's, when the development of LNG technology made possible imports from Alaska and Brunei. Domestic production is still very small. In Australia and New Zealand gas consumption is very small but significant offshore discoveries made in the past decade will help stimulate demand.

In the Less Developed Countries (LDC's), there has been limited development of natural gas production. Lack of markets for natural gas has discouraged the construction of gathering and distribution facilities. OPEC countries with large reserves currently have very small gas requirements. There have been some minor exports from Algeria and Libya as well as from the non-OPEC LDC's—Afghanistan, Argentina and Brunei.

Natural Gas Supply and Demand 1972-2000

Method of Analysis

Our projections for natural gas demand and supply in 1985 and 2000 are made on a national basis. National estimates of gas imports needed to satisfy demand are then combined by region: North America, Japan, Western Europe and the rest of WOCA. Desired gas imports may appear reasonable on a national basis, but global aggregation is needed to determine whether desired imports for all countries can be met. The total of these regional imports is an estimate of the imports required from OPEC and Communist areas—the two regions with potential surpluses.

North America

In 1975, natural gas production in the U.S.A. was 9.5 MBDOE. Since 1966, annual production of natural gas from the lower 48 states has been greater than new additions to reserves, resulting in a decline in the Reserve to Production ratio (R/P ratio) from about 17 to 1 in 1966 to about 11 to 1 in 1974. Even if the large discoveries in Alaska are added, consumption for more than a decade has been increasing appreciably faster than additions to reserves.

For conservation and technical reasons we believe that the R/P ratio of 11 to 1 will be maintained. Therefore, potential production is expected to be constrained by the volume of new discoveries. It is generally agreed that future gas discoveries are likely to be found either in deep onshore wells (below 4,500 meters), in offshore structures, or in Arctic frontier areas such as Alaska.

The rate of offshore continental shelf leasing and decontrolling the price of natural gas will affect the level of exploration and probable discovery in the U.S.A. Potential U.S.A. production in 1985 could be as high as 9 MBDOE if: 1) prices of newly discovered natural gas are deregulated in the near future; 2) the offshore continental shelf leasing program attracts many bidders for leases; and 3) these two actions result in large increases and success in exploration. Alternatively, a continuing low regulated gas price could result in gas supply falling to 7.6 MBDOE by 1985. After 1985, we assume that reserves of gas will be increasingly hard to find, and in all our cases, production declines. During the period 1985 to 2000, an increasing part of production is expected to come from Alaska (either by pipeline or LNG shipments) as the lower 48 states' production continues to decline. By 2000, U.S.A. natural gas production is expected to fall to between 5.4 and 7.2 MBDOE.

With U.S. natural gas supplies likely to decline between now and 2000, an increase in imports will be needed to maintain present consumption levels. U.S. natural gas demand projections range from 8.6 to 10 MBDOE in 1985. By 2000, demand is expected to fall to between 7.8 and 8.8 MBDOE. At such levels, maximum import requirements for the U.S.A. would be about 1.2 MBDOE in 1985, and about 2.5 MBDOE in 2000.

There are plans to import quantities of LNG. Annex 4-1 shows that these potential LNG projects for 1985 for the U.S. amount to 1.2 MBDOE. This is sufficient to meet the required U.S. gas imports for the WAES scenario cases in 1985.

In 1975, Canadian natural gas production was 1.6 MBDOE, of which about 40% was exported to the United States. Since 1971, Canadian annual production has been greater than additions to reserves with the consequence that the R/P ratio has declined from 29 to 1 in 1971 to about 24 to 1 in 1975. Proven gas reserves in Canada are extensive, amounting to some 10 billion barrels oil equivalent in accessible producing areas. Another 3 billion barrels oil equivalent have been found in the frontier areas of the Arctic and in hostile environments off the east coast. It is assumed that most future discoveries will probably be in similar areas. Production from such areas requires large reserves and large amounts of capital. Significant production by 2000 from these areas will probably require an increase in energy prices above the WAES assumptions.

Because of these factors, Canadian production is expected to decline slightly by 1985 to be between 1.1 and 1.3 MBDOE. Canadian demand is expected to increase to 1985. This increase will probably be met by reducing exports to the U.S.A. Between 1985 and 2000, production is expected to decline to between .7 and .85 MBDOE, while gas demand increases. Any increased demand will have to be satisfied by imports, and by 2000 WAES analysis shows that Canadian import requirements could be as high as 1 MBDOE.

Western Europe

Three countries in Western Europe—Norway, The Netherlands and the U.K.—have large reserves of natural gas. Gas production during the next 25 years is expected to come principally from these reserves although there may be some production in Western Germany, France and Italy (totaling about .6 MBDOE in 1985 and .4 MBDOE in 2000).

The gas industry in The Netherlands is largely based on the Groningen field, although smaller reserves of gas have been found in the Dutch areas of the North Sea. Current production from Groningen is about 1.6 MBDOE, of which about 50% is exported. Government policy is to use this field as a strategic energy reserve. Production will be allowed to peak in 1978. As it begins to decline, exports will be phased out, particularly after 1985 when existing contracts expire. Such a policy will enable existing Dutch gas reserves to fulfill domestic demands into the 21st century.

Natural gas production in the U.K. started in the mid-1960's with discoveries in the southern North Sea. These fields are now within a few years of peak production, but discoveries of gas in the northern North Sea—often in association with oil—will allow production to continue to increase into the late 1980's. Production is likely to be restricted to meet U.K. demand. It is probable that the U.K. will be a net importer of gas, since it has already contracted to import gas from the Frigg field in the Norwegian sector of the North Sea.

Natural gas production in Norway has not started, although the proven reserves could support rapid development. Norway has only a limited present need for natural gas and is expected to limit its natural gas exports for domestic fiscal policy reasons. Total production may reach .5 MBDOE in 1985 and increase to about .6 by 2000.

In Western Europe as a whole, natural gas production is expected to increase from 2.1 MBDOE in 1972 to around 3.5 MBDOE in 1985, and decline to around 2 MBDOE by 2000. We project demand to increase from about 2.2 MBDOE in 1972 to 5 MBDOE in 1985, and to between 4.6 and 6 MBDOE in year 2000. This would produce a regional shortfall, which would have to be met with imports of 1.3 MBDOE in 1985 and as much as 4 MBDOE in 2000. The 1985 import requirements can be filled with firmly contracted imports of pipeline gas from the U.S.S.R. and potential LNG and pipeline shipments from Algeria, Iran and Libya. Additional import capacity will have to be built if year 2000 demands are to be met.

Japan

Japan's interest in increasing imports of LNG stems from its desire to reduce atmospheric pollution in the large towns and cities. Japanese power companies and gas utilities have already entered into long-term LNG contracts and are expected to negotiate a number of others.

Domestic natural gas production in Japan is expected to be small—less than .1 MBDOE by the year 2000. Demand for gas, however, is expected to increase from around .15 MBDOE in 1975 to 1.5 MBDOE by the year 2000. Almost all of this demand is expected to be met through LNG imports. The maximum planned LNG import by 1985 is 1.2 MBDOE, which is more than enough to meet demand. By 2000 another .3 MBDOE capacity must be added to meet desired demand.

OPEC

Proven reserves of natural gas in OPEC are estimated to be 140 billion BOE—about 55% of WOCA reserves. Much of this is associated gas—a product of oil production which must now be wasted for lack of markets. Currently there is only a small demand within OPEC countries, either for local consumption or for reinjection to maintain oil reservoir pressure. As much as 2 to 3 MBDOE of this gas is now burned off (flared) in OPEC countries, but reinjection of substantial quantities into oil reservoirs is planned for the future.

It is expected that OPEC countries will use an increasing part of their gas for domestic purposes. The government of Iran, for example, recently announced its intent to increase natural gas as a share of domestic energy from 18% to 24% by 1978 and 35% by 1987. Other OPEC countries may follow a similar policy so that by year 2000, natural gas might contribute 30 to 40% of OPEC's internal primary energy use. If this happens, natural gas consumption within OPEC countries in the year 2000 could be as high as 5 MBDOE. Proven OPEC reserves are already large and will almost certainly be increased. Despite increases in domestic use, gas reserves could allow substantial exports. The amount depends on the policy of various OPEC countries toward the production of natural gas for export.

Figure 4-1 Natural Gas Supply and Demand in WOCA: (Case C-1)

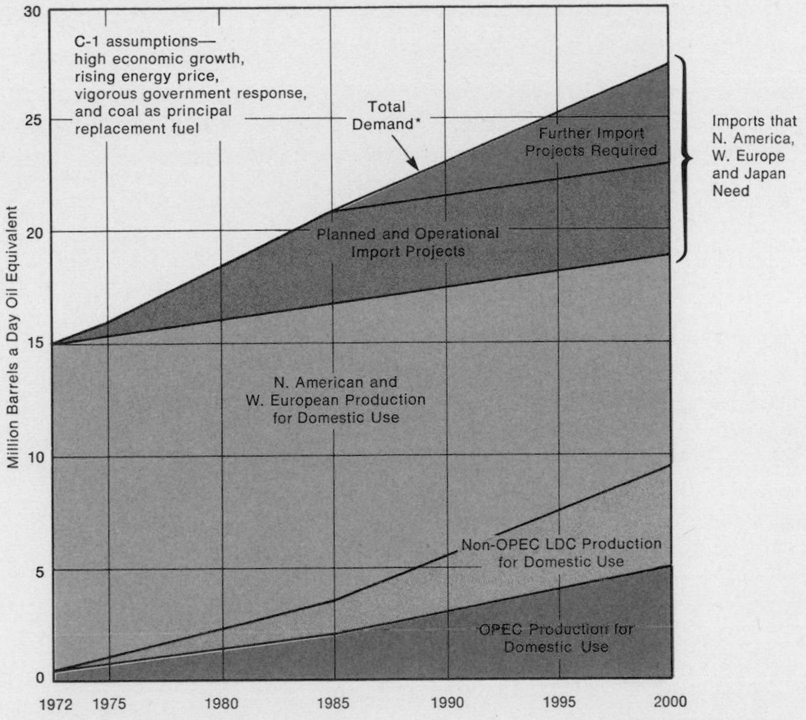

* delivered to consumer

Rest of WOCA (Non-OPEC)

Although gas consumption in 1975 was only 1 MBDOE in the rest of WOCA outside OPEC, there are estimated to be 28 billion barrels oil equivalent of proved reserves in this region. Consumption is likely to increase as reserves are developed, as has already occurred in Australia and New Zealand. By 1985, total production in the rest of the non-OPEC, WOCA countries might reach 2 to 3 MBDOE, and 3 to 4.5 MBDOE by year 2000. Most of this gas will probably be consumed locally. These countries are unlikely to import significant amounts of natural gas because of the high costs of building a gas distribution infrastructure.

Required Gas Imports in 1985 and 2000

In Table 4-5 is a summary of natural gas supply, demand and required imports for the major consuming regions of North America,

Western Europe and Japan. By 1985, import requirements amount to about 3.3 MBDOE and are fairly evenly divided between the regions. By 2000, import requirements increase to between 7.6 and 8.4 MBDOE. These figures are shown visually in Figures 4-1 and 4-2.

Table 4-5 Domestic Supply, Demand, and Import Requirements for North America, Western Europe, and Japan: 1985 and 2000 (in MBDOE)

	1985			*2000*		
	Domestic Supply	*Domestic Demand*	*Import Require-ments*	*Domestic Supply*	*Domestic Demand*	*Import Require-ments*
Case C-1						
North America	10.3	11.3	1.0	8.0	10.8	2.8
Western Europe	3.5	5.0	1.5	2.0	6.1	4.1
Japan	.1	.9	.8	.1	1.6	1.5
Total	13.9	17.2	3.3	10.1	18.5	8.4
Case D-8						
North America	8.7	9.6	.9	6.2	9.6	3.4
Western Europe	3.4	4.8	1.4	1.9	4.6	2.7
Japan	.1	.9	.8	.1	1.6	1.5
Total	12.2	15.3	3.1	8.2	15.8	7.6

Figure 4-2 Natural Gas Supply and Demand in WOCA: (Case D-8)

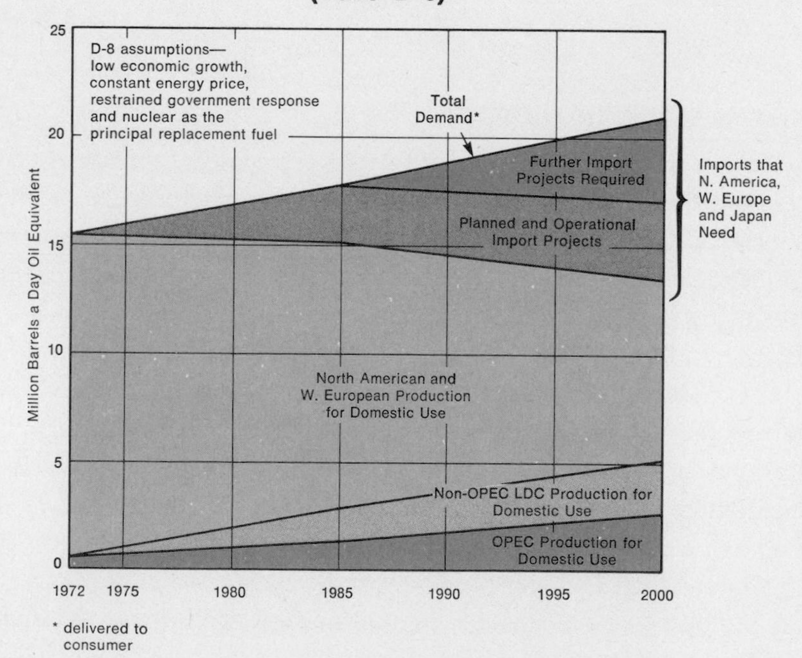

Potential Natural Gas Trade

Although international trade is currently small, there are plans to expand it which can be divided into three groups: 1) exports by the U.S.S.R.; 2) LNG exports—mainly from OPEC countries; and 3) pipeline exports from the OPEC countries in the Middle East and North Africa.

Exports from the U.S.S.R.

Table 4-6 shows existing and planned import projects for gas pipelines connecting Western Europe with the Soviet Union by

Table 4-6 Current and Potential Imports from the U.S.S.R. Into Western Europe (units = barrels a day oil equivalent)

	Markets	*1975*	*1985*
Firm Projects (including supplies under contract/option)			
	Austria	36,000	36,000
	France	—	71,000
	Finland	18,000	36,000
	Italy	36,000	142,000
	W. Germany	53,000	178,000
	Total	143,000	463,000
Possible Additional Imports			
	Austria		36,000
	Switzerland		18,000
	Total		54,000

1985. Projects in operation and under contract would allow imports of .46 MBDOE into Western Europe by 1985. Additional new projects now under discussion could, if they went ahead and were completed in time, add .06 MBDOE, making a total of .52 MBDOE by 1985. Projects for export of up to .3 MBDOE of Russian natural gas to the U.S.A. and Siberian gas to Japan are also under discussion.

Although the U.S.S.R. has large resources of natural gas which might allow larger exports, increasing domestic demand will probably limit further expansion. It is unlikely that further projects which have been under discussion will be completed before 1985. We have therefore assumed that total gas exports from the U.S.S.R. in 1985 will be .5 MBDOE. This includes projects now operating or under contract and allows for the completion of some of the projects now being discussed. For 2000, we have assumed an import figure from the U.S.S.R. of 1.0

MBDOE. This is based on an assumption that projects similar in size to all those expected to be onstream by 1985 will be completed between then and 2000.

LNG Trade

Current and planned LNG projects to Western Europe, U.S.A. and Japan are summarized in Table 4-7 and shown in detail in Annex 4-1.

Table 4-7 Current and Potential LNG Imports by 1985 (units = MBDOE)

	Current	*Under Construction*	*Further Potential*	*Total*
U.S.A.	.01	.4	.8	1.21
Western Europe	.15	.06	.57	.78
Japan	.15	.26	.76	1.17
Total	.31	.72	2.13	3.16

Source: British Petroleum, *LNG in the Next Ten Years,* 1976.

The table shows that LNG trade, starting from a very small base in 1975, could reach 3.1 MBDOE by 1985 if all the current and potential projects are completed. This would mean a tenfold increase over 1975. Projects currently under construction would furnish only about 25% of that increase. The remainder would have to come from projects classified as potential—many of which are only in the early stages of planning. About 80% of such LNG imports would come from OPEC countries.

An LNG system capable of delivering 3.1 MBDOE from OPEC to Europe, North America and Japan by 1985 would require about 25 treating and liquefaction plants, 140 LNG tankers, and 25 receiving terminals and regasification plants. Estimated capital cost for such a system would be approximately $29 billion (1975 U.S. dollars).

Pipeline Gas from OPEC

In addition, there are plans to import gas into Western Europe by pipeline from the Middle East and Northern Africa to Western Europe. By 1985, if planned projects are completed, between .2 and .3 MBDOE of natural gas could be imported by pipeline from the

Middle East and North Africa. Iran is also expected to export around .2 MBDOE of natural gas to the U.S.S.R. This may be balanced by Soviet exports to Western Europe.

Balancing Natural Gas Supply and Demand in 1985

Table 4-8 summarizes potential natural gas imports into the U.S.A., Western Europe and Japan by 1985 on the basis of firm projects and others now under discussion.

Table 4-8 Summary of Potential International Trade in Natural Gas: 1985

Exports from	
U.S.S.R. (pipeline)	.5 MBDOE
OPEC Countries	
pipeline	.5
LNG	2.5
Other Countries (LNG)	.6
Total	4.1 MBDOE

WAES projections of demand indicate required natural gas imports of 3.3 MBDOE in the main consuming regions. Realization of amount of exports shown in Table 4-8 would require the majority of potential projects to be completed by 1985.

Many projects, especially those for LNG, have not proceeded beyond the planning stage. Some projects may be cancelled, others may be delayed beyond 1985. Projects now in operation or under construction could supply about 1.0 MBDOE out of the up to 3.3 MBDOE needed in 1985. The maximum shortfall would therefore be about 2.3 MBDOE.

Balancing Supply and Demand in 2000

WAES projections indicate that to satisfy the desired demands of consumers in North America, Western Europe and Japan would require imports of 7.6 to 8.4 MBDOE of natural gas by year 2000. Imports from the U.S.S.R. by year 2000 might be 1.0 MBDOE. The remainder—6.6 to 7.4 MBDOE—is the amount of gas imports needed from the rest of WOCA. Projects currently planned and in operation could supply about half of this requirement. Additional projects would be needed to supply the remainder. Only the OPEC countries now have the reserves to supply such volumes of gas.

What is the outlook for such gas imports? Some gas might be moved by pipeline to Western Europe but most would have to be converted to LNG. Assuming that losses between the wellhead and consumer average about 25%, imports of 6.6 to 7.4 MBDOE of gas would require 8.8 to 10 MBDOE of gas to be produced at the wellhead. Domestic consumption of OPEC countries might take 5.0 MBDOE plus an unknown amount of gas for reinjection into oil fields.

Total gas production in OPEC countries would therefore need to be in excess of 15 MBDOE by 2000. In view of the scale of currently proven reserves and possible discoveries over the next 25 years, production at 15 MBDOE is probably feasible.

In the market, gas probably can command a premium price over other fuels because of its very desirable properties. However, the price which can be paid at the wellhead may be considerably below the price of equivalent oil because of the cost of liquefaction, transport and regasification. Long-term contracts between producer and consumer, which will be needed to secure the capital needed for the LNG system, must reflect such factors. Producer income from gas production will be lower than from oil production of equivalent energy content. This could make it preferable for OPEC to make greater use of gas to replace oil for domestic use, or to reinject it into oil wells to improve recovery.

Some OPEC countries have expressed some concern about the future implications of selling large amounts of gas to the same consuming countries that are now critically dependent on them for oil. On the other hand, part of the gas is produced in association with oil. In the future, OPEC countries will undoubtedly prefer to export such excess gas rather than to continue to waste it. If the price of energy rises beyond our scenario assumptions there will be added incentives for OPEC countries to export greater quantities of gas.

From the point of view of the importing countries, uncertainty about continuity of supply and price at the well may discourage the large investments required for LNG. Investments in the LNG system delivering 7.5 MBDOE would cost approximately $70 billion (1975 U.S. dollars) if capital costs are similar to those of today. The safety of LNG systems is also a concern to some. Any accident which resulted in serious casualties and property damage could affect the continuing acceptability of LNG at other delivery points.

There are some indications of what the future may hold. Iran and Algeria have recently indicated that they are reluctant to consider further natural gas export projects before 1985. These are two of the largest OPEC producers with substantial reserves. If LNG shipments are to reach the levels we have projected by 2000, production for export will be needed from these countries. With a lead time for LNG projects of about 10 years, decisions on new projects must be made soon in order to satisfy a rising demand for gas imports after 1985.

Proven reserves of gas are sufficient to support a large expansion in international trade in gas—principally as LNG. Because of its desirable qualities as a fuel, compared with oil and coal, it may command a premium price. Uncertainties about the growth of world gas trade—perhaps to levels well above our projections—lie in: attitudes of the governments of OPEC countries toward export of gas versus domestic use including use as a chemical feedstock; availability of capital for investment in LNG systems; and possible repercussions of an LNG tanker accident.

Uncertainties about the level of LNG trade in 2000 may lead to consideration of other alternatives to moving natural gas as LNG. There are at least three other possibilities.

Methanol

One alternative is to convert natural gas into liquid methanol (alcohol). This has the advantage of making transportation much simpler and it eliminates the need to build complex LNG tankers. Methanol could be used as a blending component for gasoline, in gas turbines for electricity generation or as a general fuel for many uses. It could also be reconverted to synthetic natural gas.

There are disadvantages, however. Converting natural gas to methanol involves a 40% processing loss compared to 25% for LNG. It is also capital-intensive, although it may be more economical than LNG for distances greater than 10,000 km. Transporting methanol removes the need for regasification plants in the importing country, but the direct use of methanol requires investment in new consuming equipment. Finally, as with LNG, it requires the acceptance by OPEC countries of the need to produce natural gas for export.

Coal Gasification

Large amounts of coal could be available by the year 2000, as described in Chapter 5. Coal could provide the basis for a synthetic fuels industry for the manufacture of both gas and liquids from coal. In North America, indigenous coal could be used. Western Europe and Japan would have to import coal. Substantial production of syngas from coal could help fill the gap between desired gas demand and natural gas supply in the large gas markets. However, the capital investments for this technology are high, resulting in high-cost gas, especially compared to the low prices many consumers currently pay for natural gas. Moreover, the technology is not yet satisfactorily developed. Research, development, and demonstration projects are needed to bring this alternative to the stage of commercial operation with known costs.

Reducing Demand

A third option is to reduce natural gas demand. A first step would be a change in the pricing policy for natural gas in countries where natural gas prices are now held, by government regulation, well below the price of alternatives. Another would be to restrict natural gas to its premium uses—domestic heating, chemical feedstocks—and use in geographical areas where it is needed to reduce atmospheric pollution. In the Netherlands, for instance, even with their large reserves, the government gas policy is designed to stretch reserves into the 21st century. These policies include: 1) highest priority to small customers; 2) priority for high-grade industrial applications; 3) no new supply contracts with power stations; and 4) no growth in export sales.

Conclusions

By the year 2000, natural gas production in the major consuming countries will probably decline to about two-thirds of the 1972 level. North American production will reach a plateau and start to decline before 1985. Western Europe, owing largely to expanded production in the northern North Sea, can steadily increase production until the latter part of the 1980's when it will probably start to decline, unless reserves are much larger than now estimated.

Demand for natural gas in the major consuming countries is expected to increase, but at a fairly slow rate compared to the growth over the last 15 years. North America, Western Europe and Japan together might require total natural gas imports of about 3.3 MBDOE in 1985 and 8.4 MBDOE by the year 2000 if natural gas demands are to be met.

The resources to fill these demands are available and lie primarily in the U.S.S.R. and OPEC. Pipeline gas from the U.S.S.R. may provide as much as 1 MBDOE by the year 2000. The bulk of the remaining import requirements can only come from OPEC countries. Some countries may choose to build synfuel industries that will enable them to produce synthetic gas from imported or domestic coal. Others may decide to reduce the demand for gas—reserving gas for high-priority uses only.

Policy on natural gas will vary widely from country to country. Countries that are now large users of natural gas with a large distribution system and gas-user equipment will have difficult economic and political tradeoffs to make if they are to continue to meet demands over the next 25 years in the face of declining production from domestic gas fields. Countries will have to decide to what extent they should become dependent on imports to meet their gas requirements. The costs and risks of imported gas will have to be compared with the costs and risks of imported coal and the large capital investment and environmental impacts of converting coal to synthetic gas. Some countries will decide to reduce demand for natural gas. Large gas producers, such as some OPEC countries, will have to compare the advantages of using gas (as a fuel or for petrochemicals) versus selling gas for export as LNG. These are some of the choices which nations will face within the decade.

References:

1. Kirkby and Adams, Presentation at World Petroleum Congress, Tokyo, May 1975.
2. John D. Moody and Robert E. Geiger, "Petroleum Resources: How Much Oil and Where?", *Technology Review*, March 1975.
3. *L.N.G., The Next Ten Years*, British Petroleum, 1976.

Annex 4-1 Current & Potential LNG Projects for 1985

Japanese LNG Imports

Project / Exporter	Commencing	Length of Contract (Years)	Approx. Annual Plateau Quantities: Barrels per day Oil Equivalent	Status
Alaska	1969	15	24,100	Operational
Brunei	1972	20	124,500	Operational
Abu Dhabi	1977	20	70,300	Under Construction
Indonesia	1977	20	187,000	Under Construction
Sarawak	1981-82	20	149,000	Potential Project
Australia	1983	20	162,000	Potential Project
Indonesia	1980-81	20	62,250	Potential Project
Iran	1981-82	20	62,250	Potential Project
Qutar	1983-84	20	149,000	Potential Project
U.S.S.R.	1984-85	20	175,000	Potential Project

U.S.A. LNG Imports*

Project / Exporter	Commencing	Length of Contract (Years)	Approx. Annual Plateau Quantities: Barrels per day Oil Equivalent	Status
Algeria	1976	20	8,000	Operational
Algeria	1978	20	175,000	Under Construction
Indonesia	1979	20	100,500	Under Construction
Algeria	1977-78	22	106,400	Under Construction
Algeria	1978	20	22,000	Under Construction
Algeria	1980	20	74,300	Potential Project
Nigeria	1982-83	20	125,000	Potential Project
Nigeria	1982-83	20	125,000	Potential Project
Algeria	1981-82	20	175,000	Potential Project
Iran	1981-82	20	62,250	Potential Project
Iran	1983-84	20-25	175,000	Potential Project

* In early August 1976 the Energy Resources Council issued a policy statement which recommended that an acceptable level of LNG imports for the U.S.A. would be two trillion cubic feet per annum, i.e., about 1 MBDOE.

Western Europe LNG Imports

Project / Exporter	Commencing	Length of Contract (Years)	Approx. Annual Plateau Quantities: Barrels per day Oil Equivalent	Status
Algeria	1964	15	18,100	Operational
Algeria	1965	25	10,000	Operational
Libya	1971	20	44,200	Operational
Libya	1971	15	20,000	Operational
Algeria	1972	25	62,250	Operational
Algeria	1977	20	62,250	Under Construction
Algeria	1980-81	20	62,250	Potential Project
Algeria	1980-81	20	62,250	Potential Project
Algeria	1980-81	20	62,250	Potential Project
Iran	1983-84	20-25	174,700	Potential Project
Algeria	1983-84	20	211,000	Potential Project

Source: *LNG, The Next Ten Years*, British Petroleum, 1976.

CHAPTER 5

COAL

Coal Today — The Resource Base — Demand — Coal Supply Potential — Policy Issues in the Development of Coal Production — Synthetic Oil and Gas from Coal — Implications for Action

Coal was once the world's dominant commercial fuel. For the last decade, however, its production has been level while the production of oil has increased sharply to become the major fuel in the advanced industrialized nations. World oil production is expected to peak and decline during the next 10 to 25 years, and nations must seek alternative fuels. Once again, coal may have to sustain industrial production and economic development throughout the world.

Large increases in coal production and use would have a profound impact on people's attitudes toward coal. It would require developing new infrastructure and new technology for coal mining, processing, and use. Coal could ultimately replace existing energy systems, which are designed for oil and gas. It would force the world to confront the serious environmental issues related to extensive coal mining and coal burning.

In this chapter, we assess prospects for future coal development in the WAES scenarios. We will analyze the world coal resource base, the outlook for coal demand, and potential coal production over our time frame, 1985 to 2000. The analysis will deal specifically with such potential constraints on coal production and coal usage as environmental concerns, the need for infrastructure development, and the relative prices of alternative fuels. We will also consider synthetic fuels from coal as an alternative to the direct use of coal.

The final section outlines implications of large-scale coal development as a substitute for oil, the options available, and the choices facing national policy makers.

Coal Today

Today's leading coal producers are the major industrialized countries of the Western world and the Communist area. Figure 5-1

Figure 5-1 Historical Coal Production: 1960-1974

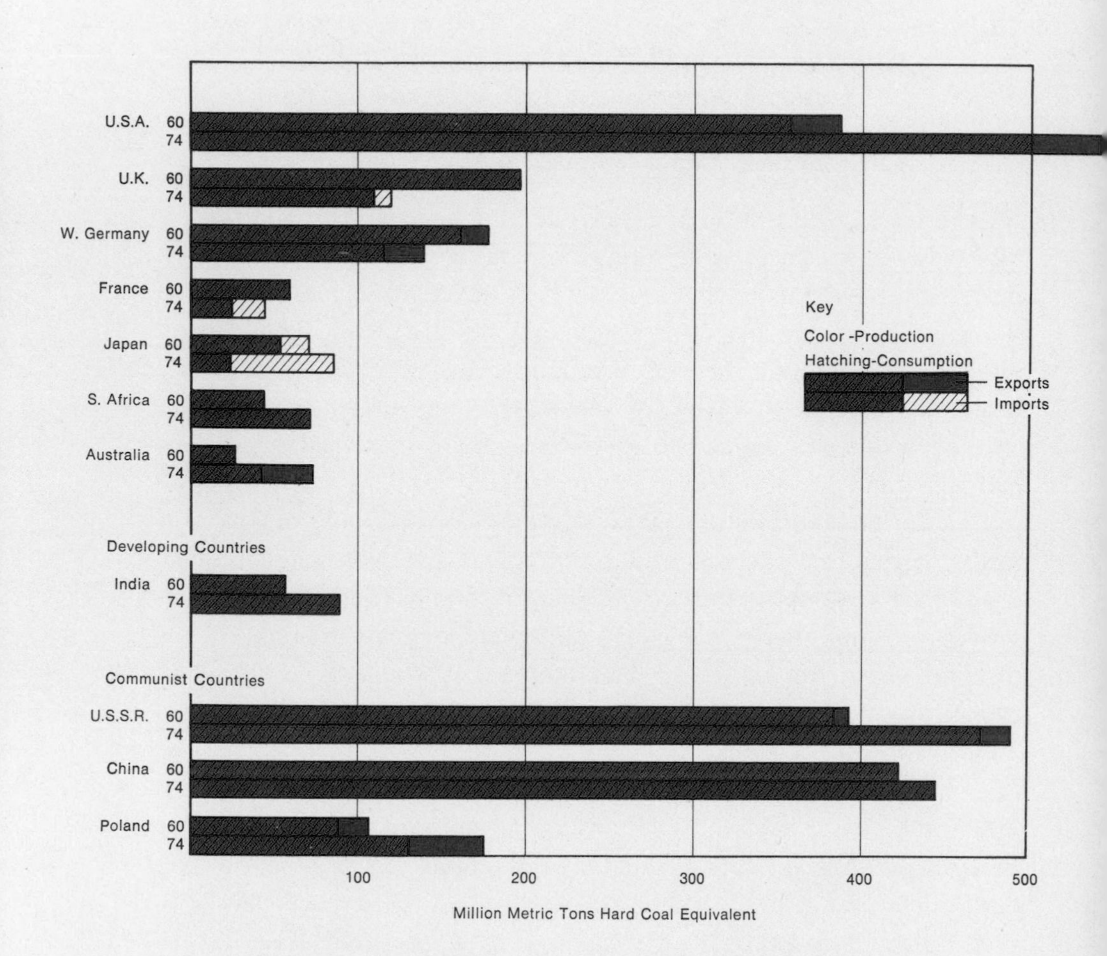

shows that three countries (U.S.A., U.S.S.R. and China) account for nearly 60% of world output, while the top six (the above plus Poland, West Germany and the United Kingdom) produce over 75%. Coal production in developing countries is similarly concentrated with India the only large producer/consumer. In WOCA, the single largest producer and consumer of coal is the United States, with nearly half the production and over half of the known reserves.

In 1974, 1,200 million metric tons of coal, or 15.2 MBDOE, provided 18% of the primary energy consumed in WOCA as shown in Table 5-1. This is a sharp decline from 34% in 1960, although total coal consumption was marginally higher in 1974 than it was in 1960. Coal use has been declining in all market sectors except electricity generation, which now accounts for half of all coal consumption.

While coal is globally a major source of energy, most coal is consumed in the country where it is mined. In 1974 only 9% of the world's production (233 out of 2,513 million metric tons) moved in international trade. A large part of this was coking coal for the iron and steel industry.

Table 5-1 Coal Consumption — 1975 (MBDOE)

	North America	*OECD Europe*	*Japan*	*Rest of WOCA*	*Total WOCA*
Coal Consumption	6.4	4.7	1.1	3.0	15.2
Total Primary Energy	37.4	23.7	6.8	15.3	83.2
Coal's Share of Primary Energy	17 %	20 %	16 %	20 %	18 %

WOCA Markets for Coal — 1975 (MBDOE)

	Rail Transport	*Industry*	*Domestic*	*Electricity Generation**	*Gas Manufacture**
Total Energy	1.0	19.8	17.6	22.7	1.0
Coal	0.3	5.2	1.2	7.5	0.8
Coal Share	35 %	26 %	7 %	33 %	76 %

* Input

The Resource Base

The world possesses vast reserves of coal, far in excess of those of any other fossil fuel. This resource base is sufficient to support massive development of coal well into the next century. The amount of coal actually produced, however, will be governed by the level of coal demand, the development of facilities for transportation, handling, storage and use and the resolution of environmental, social and economic policy issues related to coal production and use.

The World Energy Conference *Survey of Energy Resources 1974* estimates the world's *total resources*[a] of all ranks[b] of coal to be about 11,000 billion metric tons. Of this total, about 1,300 billion count as *known (measured) reserves,*[c] of which approximately 50% were deemed *economically recoverable*[d] at the time the resource studies or surveys were prepared (generally 1972 or 1973).

Table 5-2 indicates the levels of economically recoverable and measured reserves adopted for WAES national studies. Data for non-WAES areas were derived from the World Energy Conference survey.

Estimates of ultimate recoverability vary from one coal field to another, depending upon the accessibility of the coal, which is affected by a number of factors such as seam thickness and depth, type of terrain, and land ownership. Recovery rates vary widely—85-95% in surface mining and 25-70% in deep mines.

The aggregation of reserves on a ton-for-ton basis, as in Table 5-2, is only a rough indication of the extent, location and energy content of coal resources. The energy content of coal varies. Brown coal, for instance, contains only a third as much energy per ton as bituminous coal. It also contains a good deal of moisture, making it costly to transport. Yet, its sulphur content is generally lower than that of bituminous coal. Taking into account the inferior calorific value of the lower ranks of coal, total economically recoverable re-

---

Footnotes:

(a) The "sum of known reserves-in-place plus additional indicated and inferred resources which could exist in unexplored extensions of known deposits or in undiscovered deposits in known fuel-bearing areas."

(b) The sum of hard coal (anthracite and bituminous coals) and lower ranks (subbituminous, lignite and brown coals).

(c) The estimated quantity of coal "in-place in known deposits based on specific sample data, measurement of the deposits, and detailed knowledge of the quality or grade of the deposits including that part of total solid fuels in-place normally remaining in the ground due to extraction requirements."

(d) "The portion of the total amount of known reserves considered to be actually recoverable under current economic conditions and using current mining technology."

Table 5-2 World Coal Reserves
(Billion metric tons)

	Known (Measured) Reserves	*Economically Recoverable* Reserves*
U.S.A.	396	248
Canada	13	6
North America	409	254
W. Germany	100	16
United Kingdom	99	4
Rest of W. Europe	26	21
Western Europe	225	41
Japan	3	1
Rest of WOCA	140	53
WOCA	777	349
U.S.S.R. and E. Europe	349	287
China	201	101
Total World	1327	737

* Note: The majority of these estimates were prepared prior to the oil price increases in 1973/74: in the new era of higher energy prices, economically recoverable tonnages are likely to be higher.

serves of 737 billion tons of all ranks amount to some 600 billion tons of hard coal equivalent—sufficient for over 200 years' consumption at the current rate of coal usage.

Expressed in terms of oil equivalent, the amount of coal assessed as economically recoverable from proved reserves is equal to around 3,000 billion barrels—between four and five times the current level of proven reserves of crude oil. However, undiscovered reserves of coal could be much greater than the reserves now known.

Most estimates of the crude oil ultimately recoverable range around 2,000 billion barrels, as described in Chapter 3. The World Energy Conference estimate of total world coal resources of 11,000 billion tons is equivalent to some 50,000 billion barrels of oil. Taking a conservative view that an average of 25% is recoverable, a total ultimate production of 2,500 billion tons of coal, or 12,000 billion barrels oil equivalent, would be achievable—about six times the estimated level of recoverable oil. The relative orders of magnitude are illustrated in Figure 5-2.

The picture for coal is less clear than for oil because exploration has been less widespread and generally less intensive. Many estimates of coal reserves were made during an era when readily avail-

Figure 5-2 A Comparison of Oil and Coal Reserves (Units: Billion Barrels Oil Equivalent)

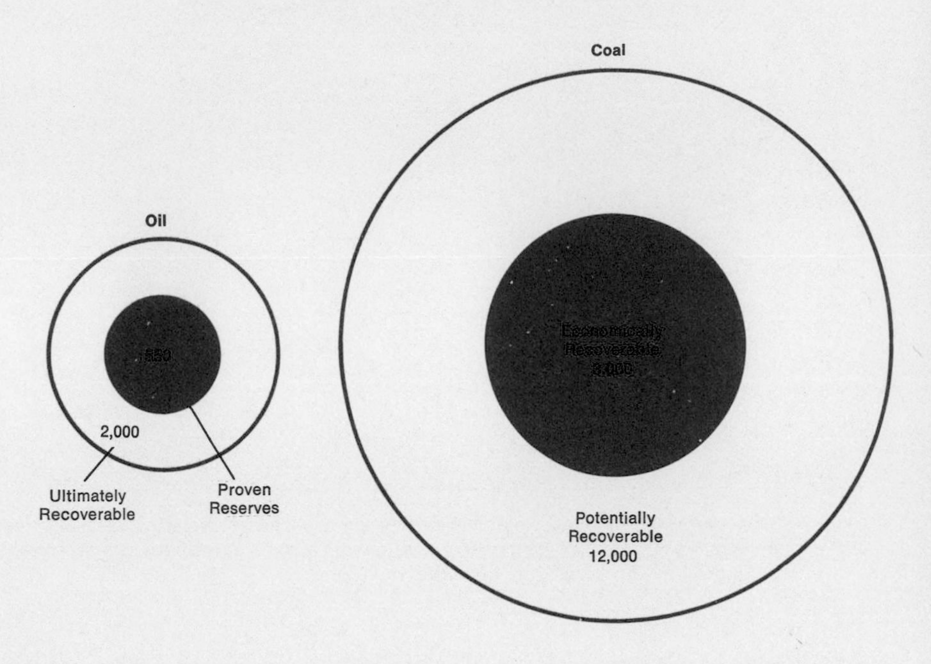

able, low-cost oil was rapidly displacing coal, resulting in little incentive to look for coal.

While coal deposits are widespread, occurring to some extent in many countries, the really large known reserves are concentrated in just three—the U.S.A., the U.S.S.R., and China. Figure 5-3 shows that much of the coal discovered so far lies in the northern temperate zone. This is perhaps because most coal exploration has been in this area. Early industrialization was based upon locally available coal and, since sufficient was found, there was no need to look further afield. Since World War II, oil and natural gas have dislodged coal from many of its former markets and have met most of the incremental demand for energy throughout the world. Few of the developing countries have ever had to look for coal. However, geology suggests that there are significant parts of the world, including the Southern Hemisphere, with a high coal potential.

Australia is a good example. Less than a hundred years ago coal was being shipped from Britain to Australia. Then an Australian

Figure 5-3 Distribution of Coal Reserves

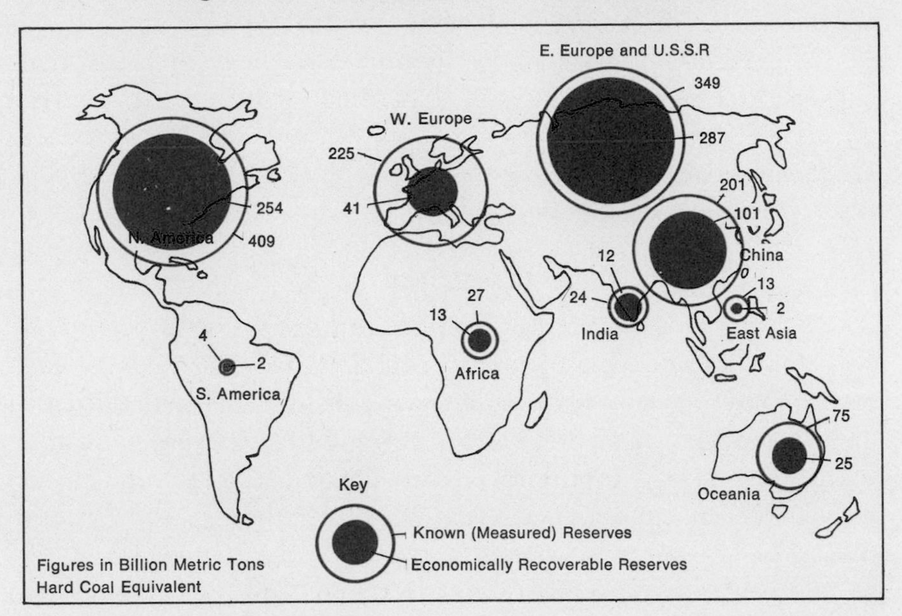

coal industry sufficient for local needs was developed, only to be dampened by oil usage. However, in the last 20 years, wider exploration has been carried out in Australia, largely to meet the Japanese demand for coking coal. As a result, output has trebled, and measured reserves have been found in New South Wales and Queensland, which, in calorific terms, are comparable with the proven oil reserves of Saudi Arabia (150 billion barrels).

Similarly, exploration in the southern part of Africa and in Indonesia is now yielding favorable results and it seems probable that the world's coal resources may be far greater than previously estimated.

But even these projections only consider coal minable from onshore locations. Just as with oil, there are believed to be major coal deposits offshore. Some of the best coal reserves in Europe are thought to lie under the North Sea. The large gas fields in the southern North Sea are in fact gas formed from coal deposits. These coal reserves are not included in even the most optimistic resource estimates.

We do not yet have the technology to mine undersea coal, but methods to exploit offshore coal would require modification of existing systems rather than invention. Alternatively, a breakthrough in in situ gasification could also change the picture, both onshore and offshore. It would be unrealistic to regard such developments as beyond future engineering capability.

Estimates of eventual coal resources can only be speculative, but the conclusion is not critically influenced by the numbers chosen. By almost any criterion, coal resources can be regarded as ample. What is in doubt is the willingness and ability of the world to accept large increases in coal production and use.

Demand

The key to estimating the future role of coal is the determination of the potential development of coal demand. The potential coal production estimated for the WAES scenarios will only be realized if the demand for coal develops and is perceived far enough in advance so that the necessary investments in production and transportation are made. The critical question, then, is whether the demand for coal will: 1) continue to stagnate; 2) rise at a rate that maintains its present percentage of primary energy supply; or 3) rise sharply as oil availability declines.

A number of conditions will have to be met before coal can again fill a growing share of the world's energy needs. People must first recognize the need for alternatives to oil and that, for future energy needs, coal is one of the most abundant and lowest-cost fuels. This implies an awareness of the longer-term energy outlook, policy decisions by governments to encourage both the production and use of coal, and public attitudes that make such policies feasible.

Widespread burning of coal will pollute the air, which could have severe environmental and climatic impacts. These issues must be faced, and ultimately resolved. If large increases in coal use are to occur, new and improved techniques for the clean burning of coal are necessary. Government policies should support research, development, and demonstration in such areas as flue-gas desulfurization, fluidized-bed combustion, and chemical coal cleaning. Governments must also settle debates on clean-air standards, and encourage study of the long-term impacts of fossil fuel combustion on the earth's atmosphere.

Extensive facilities for handling and burning coal will have to be built. This will require large investments and the development of better techniques for handling coal in a clean and convenient fashion. This is especially important in the case of industrial, commercial, and domestic users, who would otherwise have an overwhelming preference for oil or gas.

In addition to the need for realistic policies and new pollution control and coal using facilities, consumers should be given an economic incentive to choose coal over other available fuels. The cost of coal should be competitive with oil and natural gas, after allowance for the costs to consumers for facilities to use coal and control pollution.

In the WAES unconstrained case (assuming oil is not limited) integrations of desired energy demand and potential energy supply, coal's share of its traditional markets does not change much. Its share of electricity generation declines, especially in the WAES "nuclear" cases, and coal's use for transportation fuel becomes insignificant.

In global integrations limited by oil supply, however, a different pattern emerges. Total WOCA energy supplies are sufficient to meet demand only if massive fuel substitutions take place, enabling available fossil fuels to be used in the most efficient manner. This involves the elimination of fossil fuels, including coal, from electricity generation to avoid the 60-70% generating losses. Oil is reserved exclusively for transportation and petrochemical feedstocks, while coal is used directly in industrial, commercial and domestic markets.

The degree to which energy consumers will be willing and able to convert to more extensive coal use is uncertain, as is the desirability of removing nearly all fuels from electricity generation. However, our analyses give a clear directional indication of the potential use of coal as a major available alternative to oil. The magnitude of the fuel switching that may be required underscores the difficulties to be faced in making the transition from oil. Individual national governments will have to choose the extent to which they will encourage or discourage the development of expanded coal consuming systems. Because of the long lead times in changing over to an energy system based on coal, decisions must be made quickly if such a transition is to be made. The choices made in the next few years are likely to set the course for the rest of the century.

Coal Supply Potential

WOCA Coal Production Estimates

Potential coal production estimates for our scenarios were prepared by each WAES national team. A special analysis of potential coal production and exports from the rest of WOCA was also undertaken. These production estimates are summarized in Figure 5-4.

Throughout the period 1975-2000, Western Europe and Japan will probably be net importers of coal, potentially on a large scale. Areas with coal export potential will be North America, parts of Latin America, South Africa, South Asia, Australia, and certain Communist countries. The eventual level of coal available for export from each area is uncertain because it depends on estimates of both supply and demand for coal.

While the rest of the WOCA region as a whole is likely to be a net exporter of coal, there will be many countries in that region which are importers of coal, obtaining it from sources within the region such as Australia. Our estimates of potential coal supply, de-

Figure 5-4 Potential Coal Production in WOCA

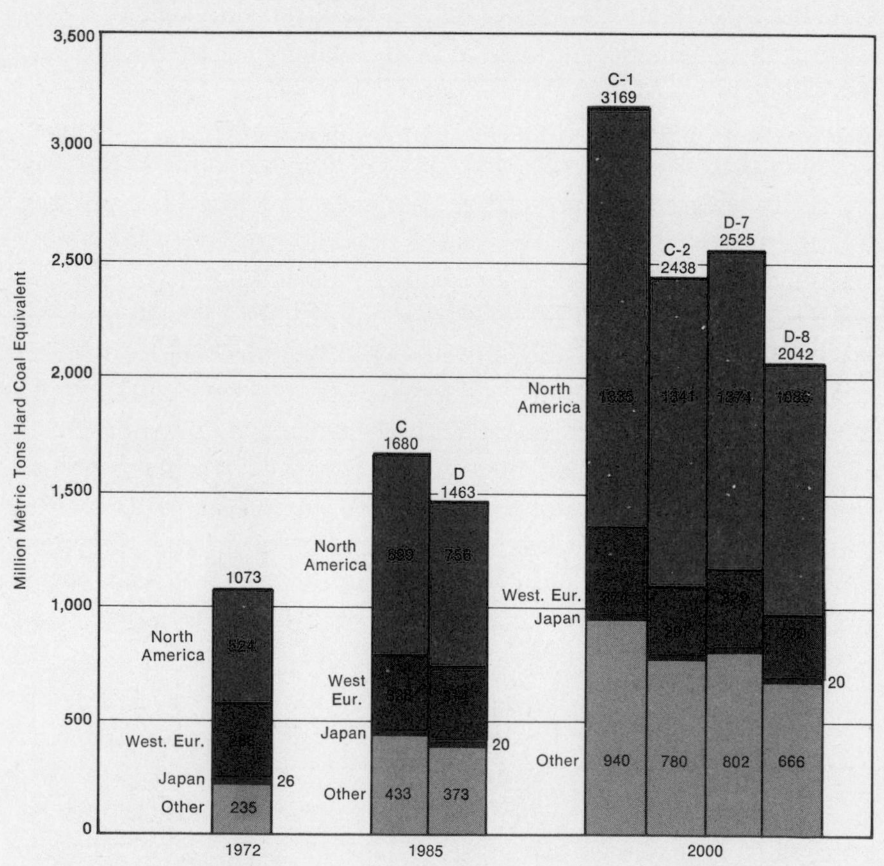

mand, and net exports from the rest of WOCA are summarized in Figure 5-5.

Of the total coal exports from the Australia/New Zealand/South Africa group, approximately two-thirds are likely to come from Australia, and the remainder from South Africa. Exports of coal from OPEC countries are expected to be largely from Indonesia, and some from Venezuela. India will probably produce about half of the non-OPEC developing countries' total coal output, possibly in the range of 200-250 million metric tons in the year 2000.

The large coal reserves in the Communist areas indicate the potential for significant coal export. WAES has not done an analysis of this potential, but assumes net export availability similar to the current levels of 40 million tons per year. These exports might be expanded if the Communist governments make the coal available and non-Communist consumers decide to buy it.

Figure 5-5 Coal Production in Rest of WOCA

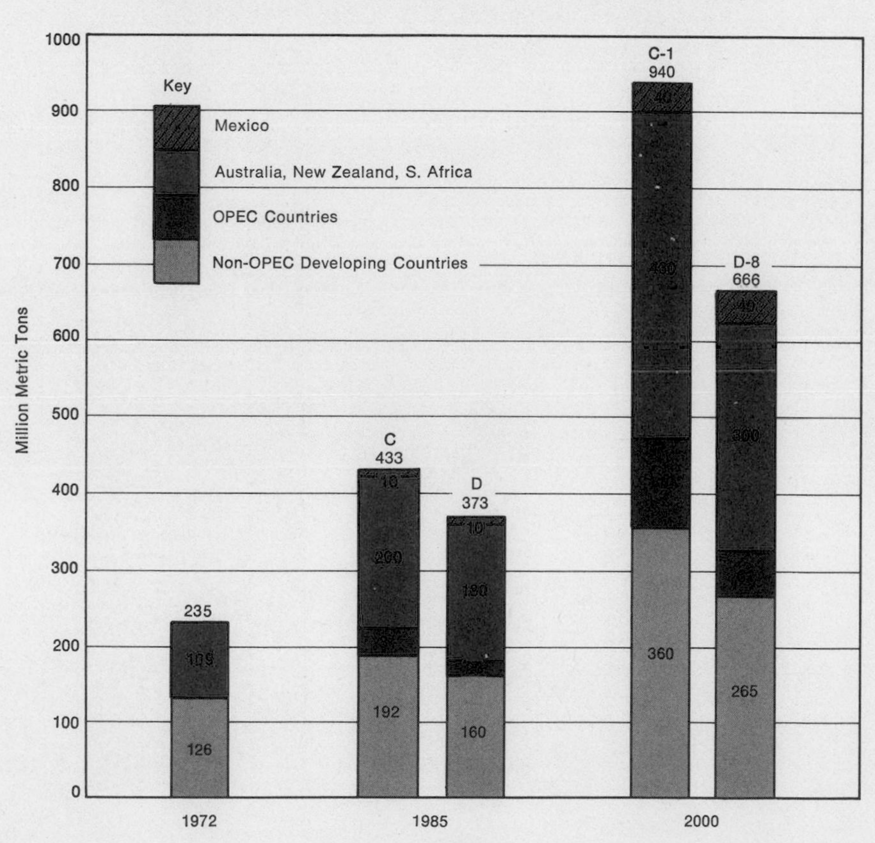

WAES did not attempt to distinguish between steam coal and metallurgical coal. Coal supply data shown in the exhibits include both steam coal and coking coal, for thermal, metallurgical or chemical purposes. The majority of incremental tonnages available for export will consist of steam coal, which is the major concern of this analysis. In the longer term, the distinction between coking coal and steam coal may become blurred, with new technology widening the types of coal usable for metallurgical purposes.

The United States as a Potential Coal Producer and Exporter

Since the United States possesses over one-half of WOCA's coal reserves and is the largest producer and consumer of coal, a special review of potential United States coal production is appropriate. Estimates of maximum potential coal development were prepared as part of the United States Supply Study for WAES, as shown in Figure 5-6, and are independent of any estimates of the demand for

Figure 5-6 Potential U.S. Coal Production

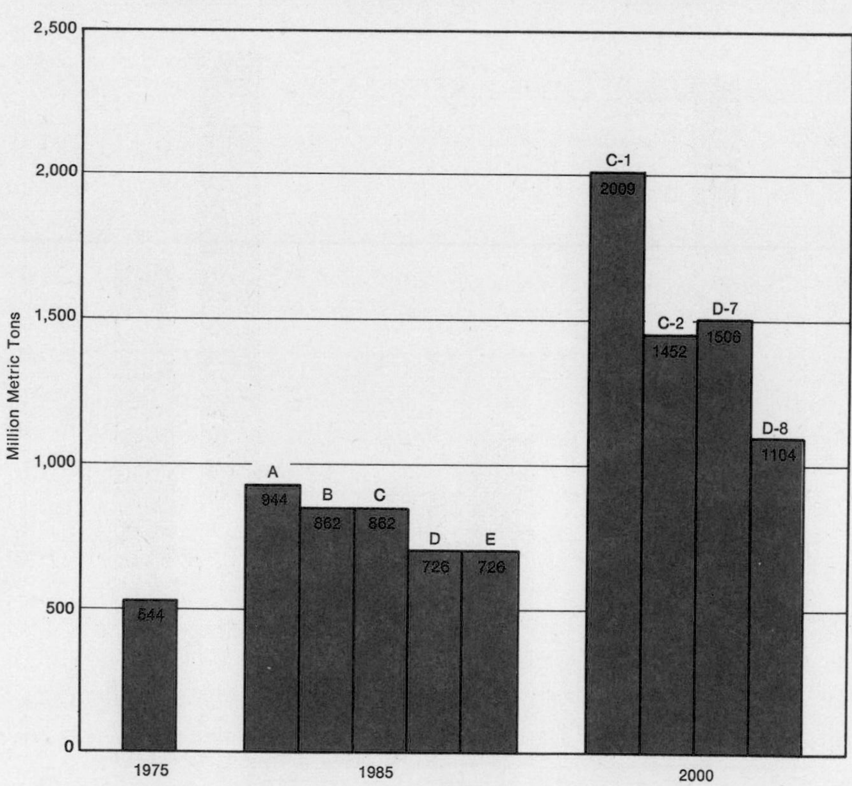

United States coal. The results of this assessment give a range of maximum *potential* coal production of 1.1 billion to 2.0 billion metric tons annually by year 2000. The low estimate assumes constant real energy prices and nuclear energy as the principal replacement fuel for oil. The high estimate assumes a rising energy price and that coal is the principal replacement fuel.

While 70% of the United States coal reserves require underground mining, it is anticipated that most new coal production will come from surface mining in the West, particularly in the Northern Great Plains Region where coal seams are thick, near the surface, and located under relatively flat land. Much of this western coal also has a low sulfur content, and can meet current emission standards for new facilities without stack gas desulfurization.

However, widespread surface mining in previously undisturbed areas is opposed by environmental groups and local agricultural interests. The two most important objections to western surface mining are: 1) the potential despoiling of large tracts of land; and 2) potentially undesirable social and economic impacts in a primarily rural, sparsely populated, agriculturally oriented society of small towns and isolated ranches. Ranchers are concerned that some of the ranch land currently used to graze livestock may be ruined for future agricultural use and that water requirements for mining and reclamation may conflict with the needs of farmers.

New mines are now being developed in the Powder River Basin of the Northern Great Plains. In a few years, these mines will be operating, and the mine developers will be in a position to demonstrate whether they can conduct mining operations in an acceptable manner and effectively rehabilitate disturbed lands. The mining companies are confident that this can be accomplished at a reasonable cost in relation to the value of the coal produced. If the acceptability of these rehabilitation programs is demonstrated, as has been done in Europe, environmental issues are not likely to be a major constraint on future coal mine development.

Recent U.S. legislation has provided assistance to local governments in planning for and meeting the cost of expanded social and community services. Most of the states involved have also adopted strict surface mining regulations that will allow new mine development, while preventing the type of abuse that occurred in the mountainous coal mining areas of the Eastern United States.

On the other hand, there is strong support in Congress for federal legislation to further regulate surface mining activities on all federal lands and limit mining in certain areas. This possible legislation and other federal/state conflicts—over regulatory jurisdiction, royalty sharing, land-use planning, and local government assistance—may slow the expansion of western surface mining activities. The significance of these issues and the related policy choices are reflected by the difference of 550 million metric tons in U.S. annual production potential between the rising energy price/coal scenario (C-1) and the rising energy price/nuclear scenario (C-2).

The high production scenarios for coal also include substantial increases in eastern underground mining in the United States. Development of this high-cost coal is consistent with the rising energy price scenarios. The relative costs of underground coal may continue to rise because of increasingly stringent mine safety regulations, associated declines in underground mining productivity, and the gradual depletion of readily accessible reserves. But development of eastern underground coal would stimulate employment and economic activity in a region of the country that has had chronic economic development and employment problems for many years.

Would the United States accept the environmental consequences and the depletion of a nonrenewable national resource to provide energy for other countries? While there is little doubt that the United States is capable of producing great quantities of coal, it is uncertain if it would be willing to undertake extensive coal mining and related expansion of transportation systems to provide coal for export.

There are a number of reasons to believe that such a policy would be adopted if the need for the coal were clearly established. Coal in the United States is not a scarce resource. Exporting it to meet the energy needs of other nations and to offset the cost of imported oil would appear to be consistent with United States foreign policy goals. Such coal development would have a positive effect on economic development, employment and the balance of payments. It will not happen, however, unless: 1) demand for coal exports is explicit in the form of firm long-term contracts, based on buyers' expectations that the coal would be available for export; 2) the environmental acceptability of western surface mining has been demonstrated; and

3) there is sufficient advanced planning to provide for infrastructure requirements, especially deep-water port facilities.

Infrastructure Requirements for Coal Development

Substantial increase in WOCA coal production will require the construction and operation of a large mining and coal transportation network to produce and move coal to market. A special analysis assessing the material, equipment, and manpower requirements for coal development in the highest U.S. coal production case was prepared for WAES by Bechtel Corporation.

The Bechtel Energy Supply Planning Model was used to calculate the resources and infrastructure needed to expand United States coal production from current annual levels (about 500 million metric tons) to 2,000 million metric tons by the year 2000 following the growth profile in Figure 5-7. The results are summarized in Tables 5-3 and 5-4.

In addition, WAES prepared a special analysis of expanded transportation facilities required for rapid expansion of U.S. coal production to 1985. A conclusion from this study was that the required transportation system to handle 400 million tons of coal from the

Figure 5-7 U.S. Coal Production Capacity (Case C-1)

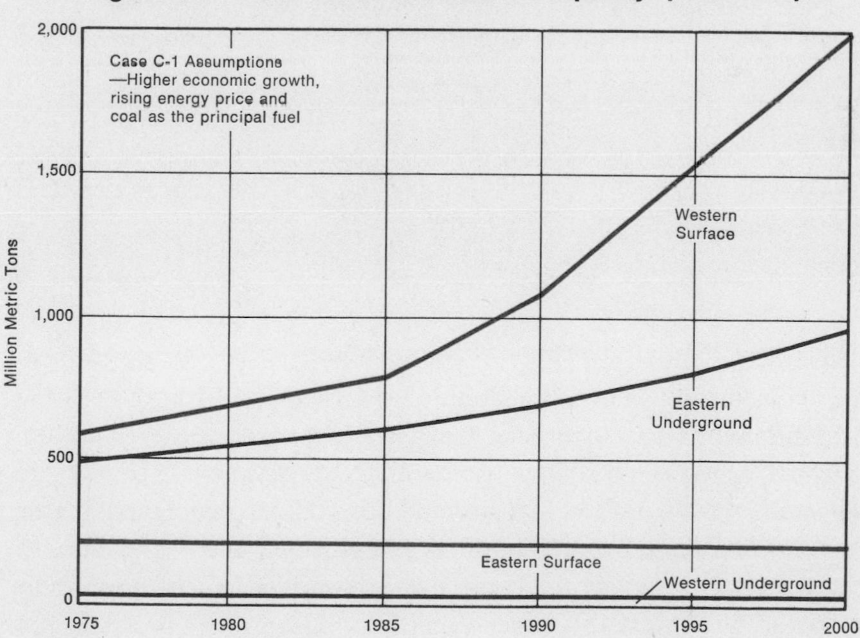

Table 5-3 Cumulative Coal Infrastructure Requirements (U.S. Coal Production Test Case)

(Production Increase 500 to 2,000 Million Tons 1975-2000)

New Coal Mine Requirements to 2000		
377	Eastern Underground Mines	— 2 Million Tons/Year
75	Eastern Surface Mines	— 4 Million Tons/Year
232	Western Surface Mines	— 6 Million Tons/Year
8	Western Underground Mines	— 2 Million Tons/Year
$32 billion	Capital for Coal Mining (constant 1975 U.S. dollars)	
New Coal-Related Transportation Facilities		
1,400	Unit Trains	
3,200	Conventional Trains	
500	Barges	
9,400	Trucks	
9	Slurry Pipelines—25 MIL Tons/Year	
$86 billion	Capital for Transportation (constant 1975 U.S. dollars)	

Source: Special study for WAES by Bechtel Corporation.

Table 5-4 Annual Requirements of Selected Resources for U.S. Coal Production Test Case

Construction	*Units*	*1975*	*2000*
Total Technical Manpower	man-years/year	7,200	22,000
Total Manual Manpower	man-years/year	14,000	42,000
Total Carbon Steel	million tons	2.1	5.2
Draglines	cubic yards	1,500	2,800
Operations			
Total Technical Manpower	man-years/year	31,000	80,000
Underground Miners	man-years/year	50,000	110,000
Total Manual Manpower	man-years/year	170,000	420,000

Source: Special study for WAES by Bechtel Corporation.

Powder River Basin can be provided by either rail or coal slurry pipeline or probably some combination of the two. Rail transport is the most suitable for local movements and for low to moderate quantities, or where maximum flexibility is required. Coal slurry pipelines are the most economical for high-volume, consistent, and long-distance transport. At the present time, the development of slurry pipeline systems in the U.S.A. faces political obstacles which would have to be resolved. The study also concluded that barges would probably not play a significant role in the movement of western coal because the barge traffic load on the Missouri-Mississippi River system is already heavy.

Although the United States is the world's leading coal exporter (56 million metric tons in 1974), port facilities are generally inadequate for any substantial expansion of coal exports. The United States has very limited capacities to accommodate the larger bulk carriers that would be most economical for large coal shipments. Additionally, coal facilities are presently concentrated in the East Coast and Great Lakes ports far from abundant western reserves. Clearly, if significant quantities of coal are to be exported, seaport expansion would be required.

The infrastructure requirements are large, but are not insurmountable, given the time period involved and the 6% maximum annual growth rate for U.S. coal production in the WAES scenarios. But any delays in the early stages of development will make it more difficult, if not impossible, to achieve the levels of U.S. coal production indicated in the WAES projections by the end of the century.

Policy Issues in the Development of Coal Production

Coal fueled the Industrial Revolution in Europe and North America and was the main industrial and domestic fuel for more than one hundred years. This long history of coal consumption has provided sophisticated coal production technology. However, underground mining is still labor-intensive, especially in Western Europe, and deep-mined coal production costs are therefore sensitive to wage inflation. While constant or rising real energy prices will continue to make underground mining economically attractive in many areas, significant increases in coal production will also require sizable development of surface mining. Productivity in surface mining is much better than in deep mines; however, it has its own problems, most notably the environmental objections to disturbing large tracts of land during mining and before restoration.

These environmental, economic and social issues are potential causes of political conflict in areas of the world that have not previously seen major coal mining. The current debate in the United States was described earlier. Whether environmental and social opposition will be a constraining factor in the expansion of coal production outside the United States is not clear. However, success or failure of U.S. efforts to reclaim surface-mined areas in the western states may have a direct impact on attitudes toward coal development in other countries with similar conditions.

Although developing countries are unlikely to follow precisely the historic pattern of industrialization based on inexpensive and plentiful energy, coal could be a major source of energy for their economic growth and development. For those countries with indigenous reserves, it provides a domestic alternative to imported oil and a potential source of export earnings. Development of coal production for export in these countries and North America could lead to a large international coal trade.

Synthetic Oil and Gas from Coal

Expansion of coal's market penetration could come from converting coal to synthetic liquid and gaseous fuels. This could be an attractive alternative to rising imports of oil or liquefied natural gas for countries that have large coal reserves. It would allow consumers to use existing oil- and gas-fired equipment, thereby putting the burden of new investments on the energy producers and processors rather than distributors and consumers. It could also simplify air-pollution control by concentrating coal use in fewer locations. However, converting coal to other forms of fuel adds to the consumers' cost of energy because of the relatively low efficiency of the conversion process (on the order of 60 to 70%) and the cost of the conversion facilities. Conversion would also mean mining significantly larger tonnages of coal and would require additional water for cooling and land reclamation.

The WAES national studies indicate rather limited potentials for the conversion of coal to synthetic fuels to the year 2000 within the WAES scenario assumptions. There are small amounts of coal gasification assumed under all scenarios, but significant development is expected only in the high-growth/high-price scenarios. Coal liquefaction is significant only in the same scenarios as a response to both higher prices and policy actions to encourage coal and synthetic fuel development. The maximum contribution of coal-based synthetics in these scenarios is estimated at 2 MBDOE, with two-thirds of the synthetic gas and virtually all of the synthetic oil produced in North America. The North American development is expected to be based on conventional technology while in Europe some high-temperature nuclear reactors may be used to gasify coal. The global quantities of synthetic products in WAES Case C-1 would require roughly 600 million metric tons of coal per year.

Such a level of synthetic fuel development will require early action to support demonstration of commercial synthetic fuel technologies. Lead times are long, but activity could begin promptly. For example, in the United States, several coal gasification projects to produce 250 million cubic feet of gas per day and costing nearly $1 billion each have been proposed and have received preliminary approvals from regulatory agencies. If favorable government decisions were taken in 1977, and financing were available, some production could begin before 1985 and could reach significant levels by the year 2000.

Because of high costs, synthetic fuels will most likely require vigorous public policy support. Synthetic gas from coal is now estimated to cost at least $17.25 per barrel oil equivalent, the energy price in the higher price WAES scenarios. The technological and economic risks associated with new processes, combined with public regulation of gas prices, may discourage private investments, although some synthetic oil is projected in the WAES rising energy price/coal scenarios, reflecting expected government assistance.

In view of the potential for large-scale conversion of coal to synthetic oil and gas, it was decided to explore the capital, manpower, equipment and material resources that would be needed to develop significant synthetic fuels production levels beyond the level assumed in the WAES scenarios. A test case was developed for the United States calling for 3 MBDOE synthetic oil and gas from coal and .5 MBD of oil from shale. It was assumed that the facilities start from zero in 1985 and reach the target production level in the year 2000 as shown in Figure 5-8. The results of the test case indicate the need for capital investments of about 90 billion dollars (U.S.) over the 20-year period from 1981 to 2000. As shown in Figure 5-9, annual increments of new facilities increase rapidly. Although manpower, equipment, and material resources required are not likely to be constraints, the projected rate of growth beginning in the early 1980's would require commitments during the next few years in order to realize the levels of production in this test case by the year 2000. The lead times for constructing specific facilities range from 3 to 5 years. The total

Figure 5-8 Synthetic Fuel Test Case Profile

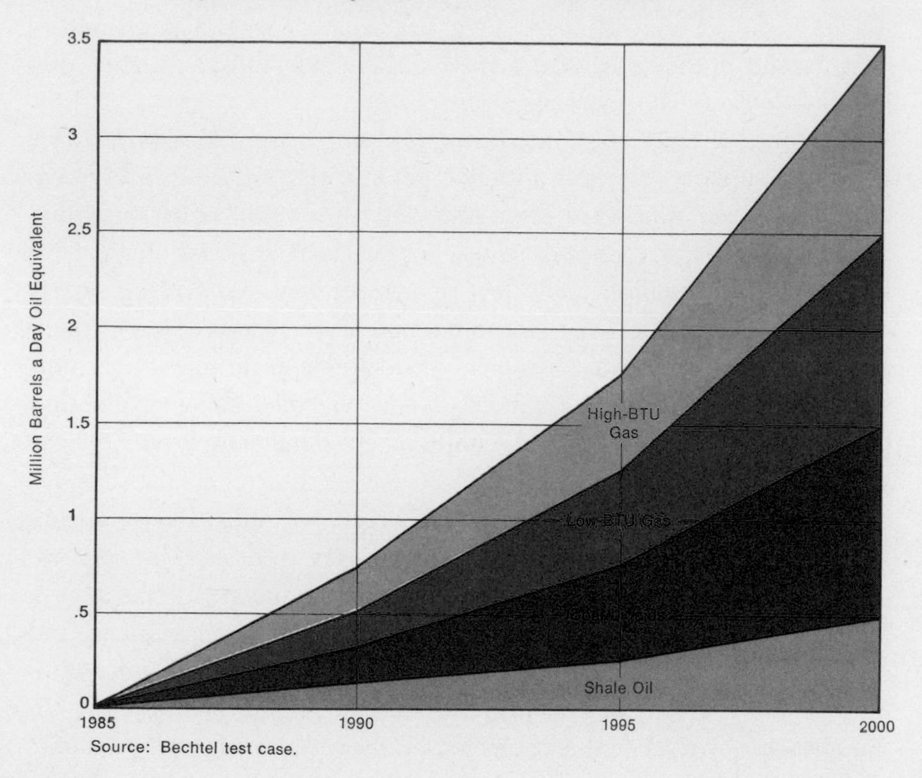

Source: Bechtel test case.

lead time—including policy decisions, investment commitments, design and engineering—stretches out from 6 to 10 years, including coal mine development and transportation. Therefore, policy decisions must be made in the next few years if facilities are to come onstream by 1985 and if there is to be significant production by the year 2000. Otherwise, synthetic fuels from coal are not likely to make an important contribution toward meeting our total energy needs in this century.

Implications for Action

Coal is the one fossil fuel that is likely to remain in abundant supply at relatively low costs for the remainder of this century and well into the next. It is one of the major alternative replacement fuels available to bridge the gap from the current era of abundant oil and gas to a future era of renewable energy resources. Coal will be even more important if energy-consuming nations are reluctant to move

Figure 5-9 Requirements for the Synthetic Fuels Test Case

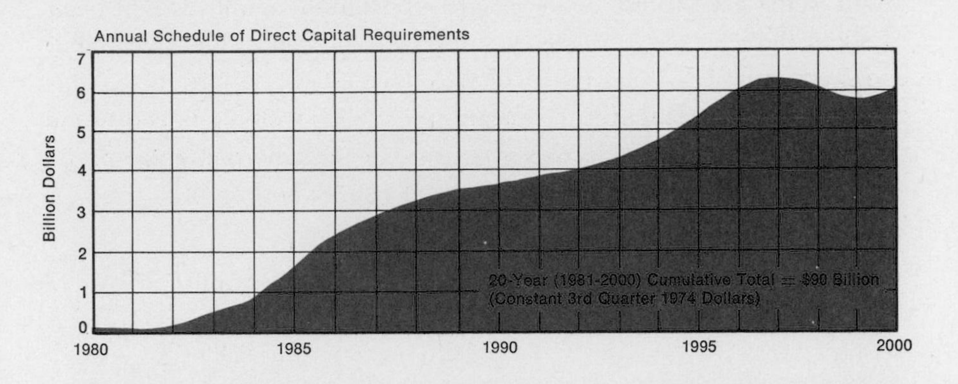

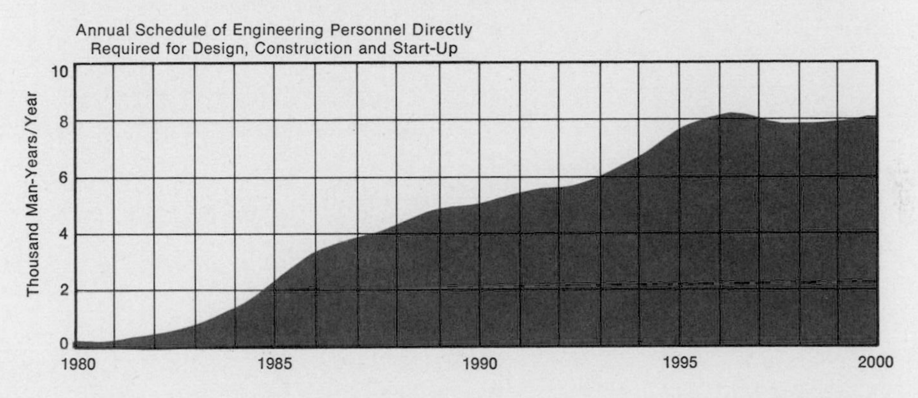

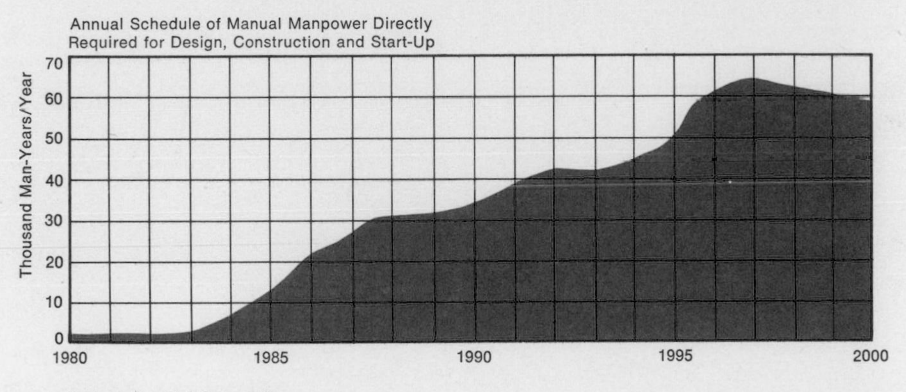

Source: Bechtel Corporation.

aggressively to implement nuclear energy. Conversely, decisions to avoid or limit the consumption of coal will put increased pressure on nuclear development and energy conservation as alternatives to unsatisfied oil demand.

Realizing coal's potential will, however, depend on a recognition of the need to use coal. Energy-consuming countries will need to adopt the necessary policies to make such use possible. Realizing coal's potential also requires favorable policies in coal-producing nations, the development of transportation systems (including deep-water port facilities), and satisfactory solutions to the environmental problems associated with mining and burning coal, wherever it is used.

CHAPTER 6

NUCLEAR ENERGY

The WAES Approach — The WAES Scenario Estimates of Future-Installed Nuclear Capacity — Some Special Characteristics of the Nuclear Safety Debate — Nuclear Fuel Cycles — Reactor Types — The WAES Nuclear Assessment — Stages of Nuclear Choice — Projections to the Year 2000 — Nuclear Fusion — The Task of WAES

The WAES Approach

Our projections of the future available supply of nuclear energy for 1985 and year 2000 are based on estimates by WAES national teams for their own countries and on our joint estimates for non-WAES Europe, OPEC and the rest of the World Outside Communist Areas (WOCA).

When we began our work in 1974, most of the WAES countries had embarked upon major programs to expand their electric power systems with large nuclear-fueled plants. Each plant of 1,000

EDITOR'S NOTE: *In the section of this chapter labeled "Nuclear Fuel Cycles," we describe the fuel cycles for the principal types of nuclear power reactors and define terms commonly used to describe nuclear systems. Some readers may wish to review this section before reading the material which follows.*

MW(e)* capacity would be large enough to supply electricity for a large city in an industrialized country. The total capacity of such plants in WOCA was projected at that time to grow from about 70,000 MW(e) in 1974 to more than 2,000,000 MW(e) by the year 2000—at an annual growth rate of about 15%—and at a cost of about $.4 to $1.0 billion (1974 U.S. dollars) per 1,000 MW(e) (excluding cost of transmission, distribution and fuel cycle facilities, which together might be more than half of the power plant cost). Such massive building programs reflected confidence that: 1) electricity use could and should grow more rapidly than other energy forms; 2) nuclear power plants were safe and could produce electricity at a lower cost than oil, gas or coal; and 3) use of nuclear energy would reduce national dependence on imports of traditional fuels. Such confidence was also based on the outstanding safety record of operation of nuclear reactors and related facilities such as enrichment and reprocessing plants over more than 25 years. Early recognition of the hazards of nuclear radiation led to adoption of stringent and costly safeguards to protect workers and the public. Strict enforcement of such safeguards, under close control by public authorities, has given this industry up to now an impressive safety record.

Yet the unique character of nuclear energy as a source of radioactivity and its potential for destruction has led to resistance in many countries to the growth of the nuclear power industry. Such resistance has focused on various concerns including reactor safety; possible diversion of nuclear materials to military or terrorist use, the safe disposal of the highly radioactive waste, and the safety of fast breeder reactors.

Proponents of nuclear energy point to the fact that radioactivity is a phenomenon long known to science, that procedures for its control and containment are well established and demonstrated and that there are other equally dangerous activities which have been accepted by society. Opponents of nuclear energy argue that it is unrealistic to expect the good safety record of the industry to be maintained in the wake of the growth and spread of nuclear capacity. The lack of a clear resolution of such differences has been a major factor in slowing down the growth of the industry from earlier expectations and is the main source of uncertainty as to its future growth.

---

* 1 MW(e) = 1 × 10^6 watts.

The WAES Scenario Estimates of Future-Installed Nuclear Capacity

In WAES, we did not attempt to seek agreement on how and when the nuclear debate would be resolved in various countries. Instead, we decided to show the scale of potential contributions to the world's primary energy needs in 1985 and year 2000 that could come from nuclear energy if an intensive nuclear power program were continued for 25 years. In this way, we could show how the pace of nuclear development would affect the total supply of primary energy required to meet the world's future energy needs. This has been our purpose in the WAES nuclear assessment.

Because nuclear power might make a substantial contribution to the world's future energy supply, we selected it as a scenario variable for our projections over the period 1985-2000. Our scenarios for this period specify a possible range of future-installed nuclear capacity for electric power generation. The low nuclear estimate is based on the assumption that coal will be the major replacement fuel for oil; the high estimate assumes that nuclear power is the principal replacement fuel. Each WAES national team developed, for their country, estimates of nuclear power for 1985 and year 2000. The maximum nuclear estimate (Case C-2) for year 2000 is defined as *maximum likely* and the minimum nuclear estimate (Case D-7) is defined as *minimum likely*. These national projections reflect the national teams' best judgment as of October 1976 of the future-installed nuclear capacity in their country under the WAES scenario asumptions. The national estimates were generally based on or derived from the latest official forecasts. Nuclear projections appear to change fairly often and therefore the date of October 1976 should be noted. For non-WAES countries, estimates were made by WAES Associates who were familiar with nuclear programs in these countries. In other cases we used OECD Nuclear Energy Agency/International Atomic Energy Agency estimates.

Our first nuclear estimates were made about the time of the December 1975 release of a joint report by the OECD Nuclear Energy Agency and the International Atomic Energy Agency, *Uranium—Resources, Production & Demand, including other Nuclear Fuel Cycle Data*. This very valuable study contains national estimates of future-installed nuclear capacity to the year 2000 submitted by many countries in early 1975. It also provided a framework for our technical assessment of the various parts of the nuclear fuel cycle, which we will summarize briefly later in this chapter. This study covering the entire nuclear fuel cycle was made under the direction of 30 experts from 20 countries. Because it carried the authority of the two principal international bodies best informed about the nuclear industry, it was an invaluable aid in making the assessment we have made. We have used their analyses and data—although we recognize, of course, that changes have occurred since early 1975 when the data were gathered, and changes will continue to occur. Revised estimates of nuclear power growth appearing in the IAEA Bulletin* of October 1975, for example, are comparable for the year 2000 in their high case with the WAES *maximum likely* (Case C-2). The WAES *minimum likely* figures for 2000 (Case D-7) are substantially below their lower figures. This is due partly to the WAES scenario variable of Low Nuclear/High Coal in Case D-7.

Reserves of uranium fuels, planned enrichment capacity, and planned reprocessing capacity were studied to determine whether they would be adequate to support the *maximum likely* and *minimum likely* levels of the WAES projections. A special review on the outlook for fast breeder reactors was made by a WAES team. The results of that review are reflected in our estimates in later sections of this chapter.

If the estimated maximum and minimum levels of installed nuclear capacity are achieved, and if nuclear power plants operate at 60% load factor to meet electricity demand, then the percentage of primary energy in the WAES cases in WOCA would be shown in Table 6-1.

The relationship of these nuclear energy projections to total energy needs as shown in the table above is very important. In the maximum case for the year 2000, the expected contribution of nuclear energy is 43 MBDOE—about as much as total 1975 oil production in WOCA. If public decisions delay and/or reduce the projected rapid growth of nuclear power (our estimates imply an 11-14% annual growth rate 1975-2000), these energy needs will have to be borne by other sources, or demand will have to be reduced with attendant consequences.

---

* Reference—IAEA Bulletin—Vol. 18, No. 5/6, p. 6.

Table 6-1 Summary of WAES Scenario Estimates for Future-Installed Nuclear Power Capacity (WOCA)

	1974	1985		2000	
	Installed Capacities	Maximum Likely	Minimum Likely	Maximum Likely	Minimum Likely
Nuclear Capacity GW(e)[1]	66.9	412	291	1772	913
Percent of Primary Energy in WOCA	2	9	6	21	14
Oil Equivalent (MBDOE)[2]	1.7	10	7	43	22

[1] 1 GW(e) = 1,000 Megawatts (electric) = 1 x 10^9 watts (installed capacity) for a power plant.

[2] We have adopted the convention of expressing nuclear electricity production in terms of the fuel input that would be required to produce the equivalent amount of electricity output from fossil-fueled power stations. The assumed generating efficiencies are 35% in 2000. The formula used to convert GW(e) to MBDOE, assuming 35% efficiency and a 60% load factor is:

$$GW(e) \times \frac{8760}{620,000} \times \frac{1}{.35} \times .6 = \text{MBDOE primary energy input}$$

In other chapters, we have estimated the maximum supply potentials for oil, natural gas, coal and other sources. None of these seem likely to be capable of much expansion beyond our highest estimates without much higher energy prices and/or "super-vigorous" policy actions. Therefore, shortfalls in nuclear power would most likely increase the prospective shortage between desired fuels and probable supply. Answers are needed to the following questions: How likely is a shortage in the WAES nuclear power projections? Would it occur in some countries but not in others? What kinds of uncertainties may affect the realization of the WAES nuclear projections?

Some Special Characteristics of the Nuclear Safety Debate

As we have said, development of nuclear power is being carried forward in an atmosphere of concern about the radioactivity which is the main difference between nuclear and other forms of energy. In spite of the positive safety record of reactor operation up to this time, the nuclear industry has been called upon to prove the safety of its reactors and the safety of every step in the fuel cycle. The question of safety has been debated publicly and often in terms similar to those which have dominated discussions of atomic weapons and the threats that they represent to human survival.

There are several serious problems around which debate has turned. Probably the most serious is that of containment of radioactivity. The safe transport, storage and treatment of spent fuel elements and the resulting highly radioactive wastes which remain active for hundreds and thousands of years is a matter of great public concern. Further research and demonstration is required before the most acceptable processing methods and the most suitable sites for waste disposal can be selected.

Still another concern is the safety of nuclear power plants themselves, particularly the consequences of a failure which might lead to a release of radioactivity. Although it may become widely accepted that the probability of such a failure is very low, there is a fear of the unpredictable and possibly widespread effect of such an accident. The proposal to make prototype fast breeder reactors which produce heat in a small space has heightened this concern over reactor safety, despite the required introduction of special safety features in the design to provide for the possibilities of melting down of the core.

Many people are concerned about which materials are usable for nuclear weapons. The explanation given here may help the reader. Uranium for power reactor fuel at 2 to 4% enrichment in U-235 is useless for weapons. Similarly U-233 mixed with natural uranium to the level needed for reactor fuel (below 15% U-233) is useless for weapons. Such fuel grade uranium can only be made into weapon-grade materials in an enrichment plant. Uranium enriched significantly above a U-235 content of 20%, or U-233 made from thorium, or plutonium are all usable in nuclear weapons.

Regarding plutonium, a conventional chemical process can be used for obtaining materials from spent fuel elements for use in weapons, although it requires relatively complicated equipment. No means for rendering plutonium useless for weapons is known.

Nations that made nuclear weapons have first used nuclear reactors aiming only at producing plutonium; any sizable country could, without unsurmountable difficulties, obtain the same result in the near future, without resorting to nuclear power generating technology. Most nuclear power reactors can also partially use plutonium in the form of oxide as fuel, and large fast breeder reactors (FBR's) will require several tons of plutonium (or highly enriched U-235). There is public concern about this part of the nuclear fuel cycle.

Another area of concern is the "front end" of the nuclear fuel cycle where the fissionable fraction (Uranium 235) in natural uranium can be concentrated; nuclear power reactors require 2 to 4% U-235 concentration as fuel in the form of uranium oxide. Enrichment plants using the same technology can be used for producing the highly enriched U-235 for use in weapons and have been doing so for the past twenty-five years. It should be noted that this latter method for obtaining weapon-grade material is more difficult than the plutonium route mentioned above.

The public debate of such issues is widespread and goes on in many countries at different levels of intensity. There are those who argue that there should be a cessation of such nuclear activities as additions to fuel reprocessing facilities and the development of the fast breeder reactor. There are others who point out that, in spite of the extensive experience and small-scale experiment to date, these issues can only be resolved by continuing development and demonstration. The attitudes of the general public on these questions are beginning to be reflected in the political process. In some countries, political action has prevented or delayed the siting of nuclear plants. On the other hand, recent referenda in seven states in the U.S.A. in 1976 all gave a large majority of support for a qualified continuation of nuclear development.

The increasing desire of many countries, including some in the developing world, for nuclear capacity to meet their energy needs has been accompanied by a rising international concern about the spread of the capability for making weapons based on plutonium and a move towards more effective controls including the possibility of multilateral operation and control of fuel reprocessing plants and stricter international control of waste disposal.

Later in this chapter, we call attention to three stages of choice in the nuclear fuel cycle, and point out the possibility of considering separately (a) reactor operation which involves a single use of uranium without fuel reprocessing, (b) the processing of used fuel to extract and recycle plutonium and uranium, and (c) operation of fast breeder reactors.

We point out later some of the consequences of continued delay in decisions, and the effects of such uncertainties on the WAES energy supply projections to the year 2000. Here we would emphasize only that the viable nuclear industry that would be needed to supply and fuel future nuclear power plants may not survive widespread moratoria or long delays on nuclear plant construction.

Although the debate on these issues continues, our projections necessarily assume the existence of a large and vigorous global nuclear industry capable of building one or two thousand nuclear power plants over the next 25 years, along with all of the associated parts of the nuclear fuel cycle. Such an industry would have to be capable of supplying power plants and related facilities whose total cost over the 25 years would range from $1,000 billion to $3,000 billion in constant 1975 U.S. dollars.

Nuclear Fuel Cycles

Reactor Types

There are several different designs of nuclear power reactors. Each type has distinctive performance characteristics, economic considerations, fuel requirements, etc. The OECD/IAEA study, for example, projects seven kinds of reactors in their estimates of reactor types to be built by the year 2000. For analysis of fuel cycle requirements for WAES cases, we were guided generally by the OECD/IAEA projections of reactor mix, because the mix of reactor types affects requirements in stages of the fuel cycle. For 1985 we used LWR 90%, HWR 5%, other 5%. For 2000 we used LWR 80%, HWR 7%, FBR 5%, other 8%. We confine our assessment to these three principal nuclear power reactor types: the LWR, or Light Water Reactor (Pressurized Water—PWR, Boiling Water—BWR); the HWR, or Heavy Water Reactor (CANDU-type); and the Fast Breeder Reactor (FBR).

Reactor Design Factors

Nuclear plants are similar to fossil fuel plants in that both systems generate steam to turn turbines connected to generators that produce electricity. The difference is that nuclear plants use nuclear fuels.

A nuclear energy plant—a reactor—is based on controlling a chain reaction in fissile material. The fissile material—U-235—is "diluted" with other material (U-238), but the all-important fissile nuclei (U-235) are the nuclear fuel. When a neutron is absorbed into a U-235 nucleus, the nucleus splits with the release of a tremendous amount of energy which is ultimately converted into heat and also ejects 2-3 neutrons. Three things may happen to these neutrons: 1) some may go out through the surface of the uranium mass and get lost; 2) some are absorbed by a nucleus of U-238, which then becomes plutonium (Pu-239) which is a fissile material like U-235; or 3) some will be absorbed into other U-235 nuclei. At least one neutron must enter a U-235 nucleus to keep the chain reaction going. The U-235 nucleus splits, releases heat and ejects 2-3 neutrons.

The "desired reproduction factor of 1" means that from each U-235 nucleus which splits at least one neutron is absorbed into another U-235 nucleus to produce fission. When a reactor has a reproduction factor of 1 the system is "critical" and the chain reaction, which produces the heat, continues.

The system depends on a large enough quantity of uranium, the critical mass, to contain enough nuclei to maintain a chain reaction. Naturally occurring uranium (an element found in natural deposits) contains only 0.7% fissile U-235 nuclei. Enrichment increases this concentration to about 3%. This still relatively low concentration is adequate to sustain a chain reaction in a reactor provided the neutrons from fission are slowed down by a "moderator"—a material with light nuclei, such as ordinary water, heavy water (D_2O) or carbon in the form of graphite.

A nuclear reactor, then, consists of fissile material (U-235), usually in the form of rods, in association with moderators. Adjustment of control rods which are high neutron absorbers in the reactor "core" where the reaction takes place keeps the system "critical" and allows shutdown when the control rods are pushed into the core to absorb neutrons, stopping the reaction.

Some understanding of this fission phenomenon is needed to appreciate the reasons for the differences in the fuel cycles in the Light Water Reactor and the Heavy Water Reactor. These differences are especially important for the front end of the fuel cycle. Natural uranium must be enriched from 0.7% U-235 to about 3.0% U-235 to make fuel for LWR's. Present enrichment plants are very expensive and use a lot of energy. New enrichment plants based on the centrifuge process require less energy. The HWR uses natural uranium as fuel but requires a large initial stock of heavy water (D_20) and small amounts during operation to make up for losses. Plants to make heavy water are also very expensive and use a lot of energy. The "back-end" of the fuel cycles of the LWR and HWR are similar although at present it is not expected that used HWR fuel will be reprocessed to recover plutonium and uranium.

The Light Water Reactor — LWR

There are two types of LWR's—the Pressurized Water Reactor (PWR) and the Boiling Water Reactor (BWR). They use the same fuel cycle. The fuel cycle for the Light Water Reactor (LWR) (see Figure 6-1) begins with uranium ore, a natural resource that must be discovered, mined and processed to form a concentrate (U_30_8) which is called yellowcake.

The uranium concentrate (U_30_8) is then converted to a gas (uranium hexafluoride—UF_6) which goes to an enrichment plant where the concentration of U-235 is increased from 0.7% in natural uranium to about 3.0% for LWR fuel. Uranium tails, which are ura-

Figure 6-1 Fuel Cycle of Light Water Reactor

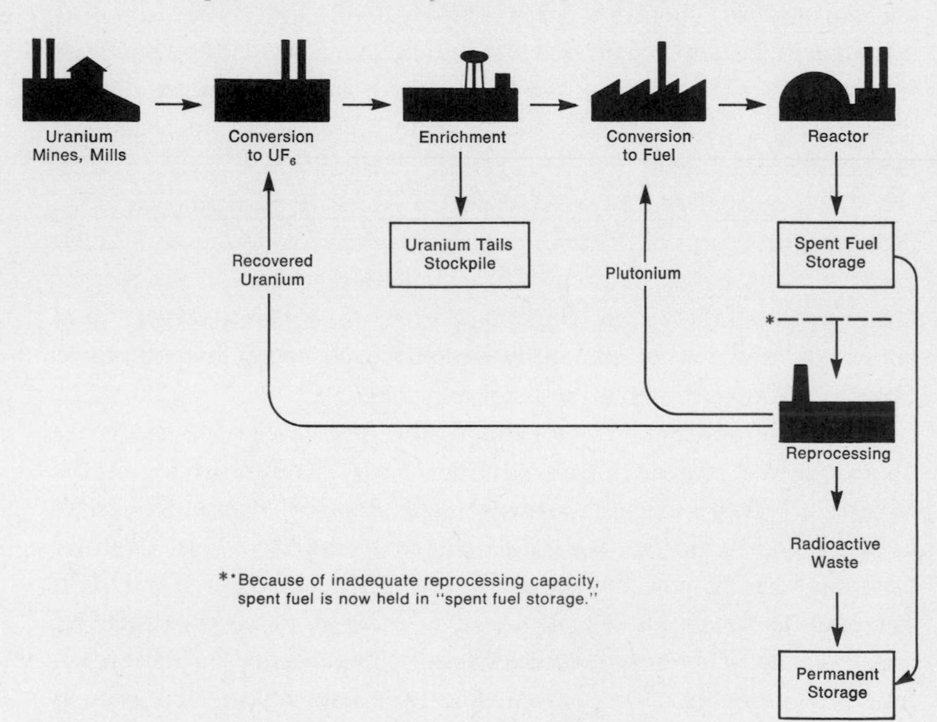

nium depleted into 0.2 or 0.3% U-235, are a by-product of the enrichment processes and are saved to use in Fast Breeder Reactors. Fabrication follows enrichment and includes converting UF_6 to uranium dioxide (UO_2) which is the final fuel, pelletizing, encapsulating in tubes, and assembling the uranium dioxide tubes into fuel elements. Finally, the fuel elements are loaded into reactors, which use the heat from nuclear fission to produce steam to drive turbines and generators to produce electricity.

The steps described above are known as the "front end" of the nuclear fuel cycle. The "back end" of the cycle begins when the used fuel elements are removed from the reactor. According to a predetermined schedule, 20-30% of the fuel is removed each year. Used fuel is highly radioactive and produces some heat because of the slow decay of the radioactive fission products. The used fuel must be stored under water at the reactor site for many months to insure removal of the heat and to shield against the radiation emitted by the fission products.

The used fuel elements can then be transported in specially shielded containers to a reprocessing plant. Here they are mechanically chopped up, dissolved in acid, and go through a chemical process to separate out three components: 1) the remaining uranium; 2) plutonium; and 3) the radioactive fission products. Such reprocessing plants are complex, because the materials to be recovered (U and Pu) are mixed with highly radioactive fission products. Although these eventually reach the point where they can be handled without shielding, many operations must be performed by remote control by operators protected by thick walls of radioactive shielding. Such plants have been operating for up to 30 years to recover Pu for weapons, so the technology for remote handling has been well developed.

At this stage there are a number of possible and important choices. The recovered uranium can be converted to UF_6 as feed for an enrichment plant or it can be stored. The uranium in reprocessed fuel from LWR's is normally richer in U-235 (0.8%) than in natural uranium. The plutonium can be recycled into the LWR for use with uranium as reactor fuel; it can be stored; or it can be used as fuel for the fast breeder reactor (described later). The radioactive waste must be stored in permanent and "leakproof" facilities. Another alternative is not to reprocess the used fuel elements but to put them directly into permanent storage repositories. This is a single-use system.

Plutonium does not occur in nature. It is a transuranic element which is a by-product of all nuclear reactor operations where U-238 is present and absorbs neutrons with the result that it is transmuted into plutonium. If recovered plutonium and uranium are recycled into reactors as fuel, they could replace about 20% of the fresh uranium required for fueling the LWR. Requirements for natural uranium could be reduced proportionately.

The Heavy Water Reactor — HWR (CANDU-type)

The Heavy Water Reactor fuel cycle (see Figure 6-2) differs substantially from that of the LWR. HWR nuclear power plants using natural uranium as fuel and heavy water (D_20) rather than light water (H_20) as the moderator have been developed and are now operating in Canada in the four CANDU units at Toronto (2.0 GW(e)) since 1972. Nuclear capacity in operation and under construction in Canada (up to and including Pickering B) now totals 8.8 GW(e). In the HWR fuel cycle, shown below, the uranium fuel is used only once.

Figure 6-2 Fuel Cycle of Heavy Water Reactor

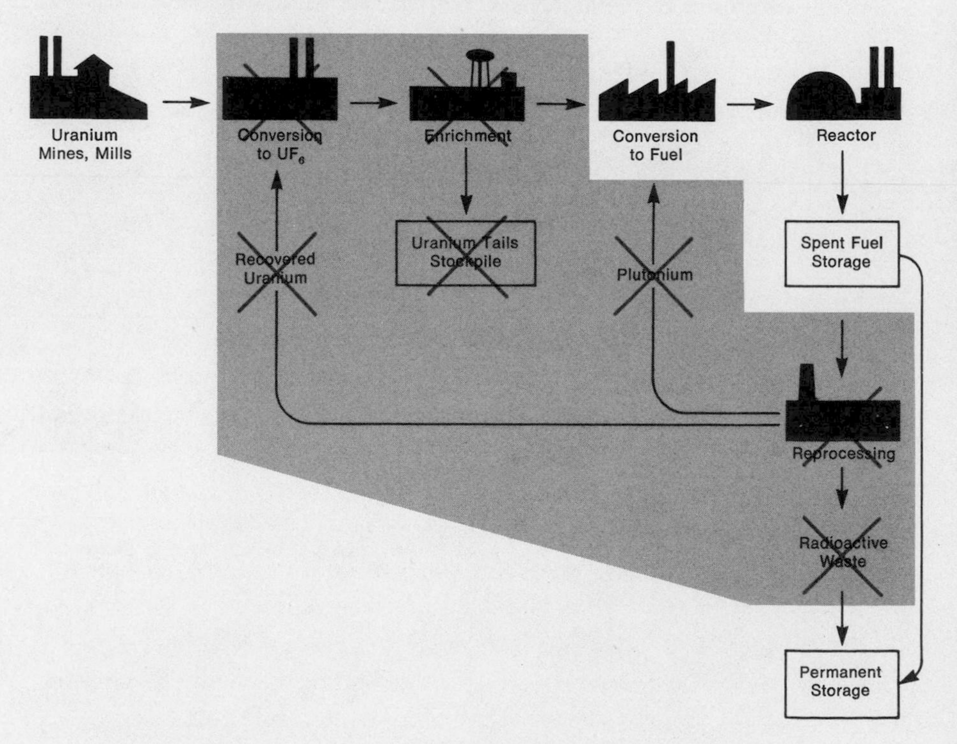

The used fuel is now stored without plans for reprocessing (single use).

The HWR cycle uses natural uranium as the fuel, thereby eliminating the conversion to UF_6, the enrichment step, and the creation of a stockpile of depleted uranium tails (0.2-0.3% U-235). In Canada today, used fuel elements are put into storage where they await future decisions as to whether they will: a) go to reprocessing operations to recover uranum and plutonium and disposal of radioactive wastes; or b) be placed in permanent storage repositories without reprocessing.

The Fast Breeder Reactor — FBR

The LWR and HWR are thermal reactors because they operate with thermal neutron energies. Heat is generated by energy released during nuclear fission. This heat is carried away by a coolant and used to raise steam. What then is a breeder reactor and how does it differ from a thermal reactor?

The Fast Breeder Reactor (see Figure 6-3) is so named be-

Figure 6-3 Fuel Cycle of Fast Breeder Reactor

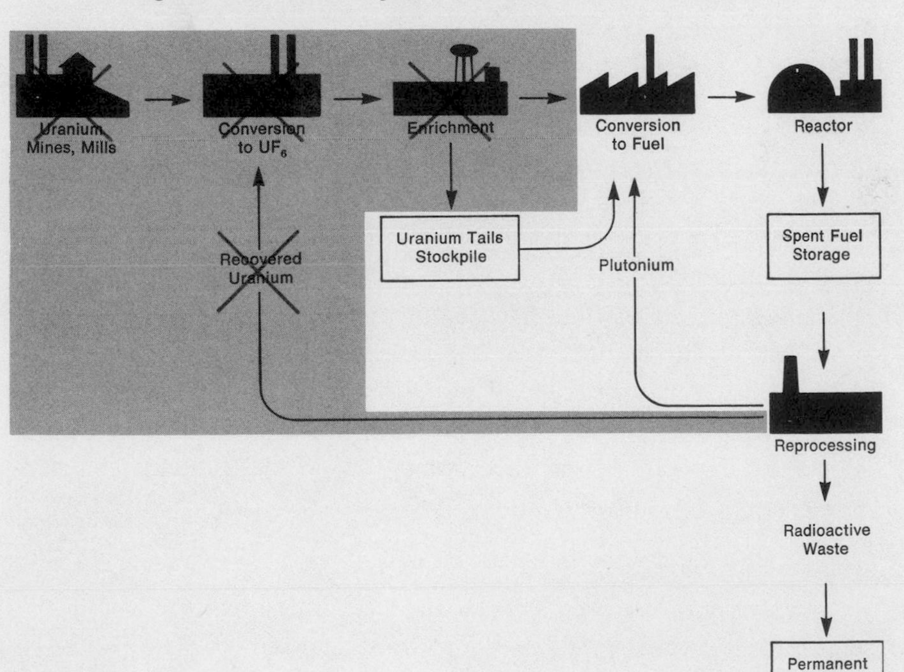

cause it operates with fast neutrons (energies $10^7 \times$ thermal) and "breeds" more fuel than it consumes. It also operates as a power plant to generate steam. It provides a way to use most of the 99% of natural uranium that is the nonfissionable U-238. The breeder converts uranium (U-238) by absorption of neutrons to produce plutonium—which is fissionable. Thus, by extracting energy from most of the uranium (U-238) that is unused by present reactors, fast breeders offer the possibility of multiplying (perhaps 50 times) the usability of natural uranium resources. The initial fuel loading of an FBR requires large amounts of plutonium which must be obtained by reprocessing fuel elements from thermal reactors.

The High Temperature Reactor — HTR

Successful development of the High Temperature Reactor could create new uses for nuclear energy because such reactor systems are expected to deliver heat at high temperatures for such industrial purposes as coal gasification and hydrogen production. Large-scale HTR prototypes have been built in Germany and in the U.S.A. Problems encountered in commercial scale units and the accompanying fuel cycle have not yet been resolved.

The WAES Nuclear Assessment

WAES Scenario Estimates of Future-Installed Nuclear Capacity

In the introduction to this chapter we explained how our estimates were developed and presented the nuclear energy share of total projected primary energy. Table 6-2 lists the actual installed nuclear capacity for 1974 for the countries and regions we examined and our estimates of future installed nuclear capacity for 1985 and year 2000. Note that the table lists projections for Mexico, a member of WAES, separately from the "Rest of WOCA."

Figure 6-4 shows the projections of installed nuclear electric generating capacity based on estimates by WAES national teams of *maximum likely* and *minimum likely* levels for 1985 and 2000.

Uranium Resources

Uranium reserves are classified as "Reasonably Assured Reserves" (RAR) and "Estimated Additional Resources" (EAR).* The

* See OECD/IAEA Report.

Table 6-2 WAES Scenario Estimates of Future-Installed Nuclear Capacity (As of October 1976)

GW(e) of Installed Nuclear Capacity

Region or Country	1974	1985		2000			
				Rising energy price High economic growth		Constant energy price Low economic growth	
	Installed Capacity*	D (Minimum Likely)	C (Maximum Likely)	C-1 "Coal"	C-2 "Nuclear" (Maximum Likely)	D-7 "Coal"	D-8 "Nuclear" (Minimum Likely)
Denmark	0	1	1	2	5	2	5
Finland	0	2.2	3.2	6.6	16.2	4.2	8.2
France	2.9	35	45	90	140	70	100
Germany	4.0	21	33	50	120	50	80
Italy	.6	5.4	9.4	20	60	20	60
The Netherlands	.5	1.5	2.5	10	16	6.5	10
Norway	0	0	0	4	4	2	3
Sweden	2.6	3.8	7.4	10	20	10	20
U.K.	5.8	14	15	33	58	29	50
Non-WAES Europe**	2.5	23	31	72	94	72	94
Total Europe	18.9	106.9	147.5	297.6	533.2	265.7	430.2
Canada	2.5	8	12	44.5	74	44.5	74
United States	40.4	127	166	380	620	380	620
Total No. America	42.9	135	178	424.5	694	424.5	694
Japan	3.9	25	40	75	120	75	100
Rest of WOCA	1.2	24	47	212	425	148	300
(including Mexico)	0	(2)	(3)	(40)	(80)	(20)	(30)
TOTAL WOCA	66.9	290.9	412.5	1009.1	1772.2	913.2	1524.2

* From The Atlantic Council, "Nuclear Policy" as revised for WAES countries.

** Non-WAES Europe includes Belgium, Luxembourg, Ireland, Austria, Greece, Yugoslavia, Portugal, Spain and Switzerland.

grades of such reserves are in the range of 0.05%-0.20% or 10-40 lbs. (5-20 kg) of U_3O_8 per ton of ore. In the OECD/IAEA Report the total of RAR was estimated to be about one million tons of U_3O_8 up to \$15/lb. U_3O_8, and an additional 730,000 tons at prices up to \$30/lb. U_3O_8. Estimated Additional Resources were one million tons up to \$15/lb. and 680,000 tons at prices up to \$30/lb. U_3O_8. Of the combined total of RAR and EAR, 72% lie in Australia, Canada, South Africa and the U.S.A.

In addition to these reserves and resources, there is a range of higher resource estimates because: 1) until now, only a fraction of the earth's crust has been explored; 2) the price increases of the last two years to levels above \$30/lb. U_3O_8 are not yet reflected in published reserve figures; and 3) uranium is known to be found in much lower concentrations which could be produced at much higher

Figure 6-4 Projected Growth of Nuclear Power in WOCA

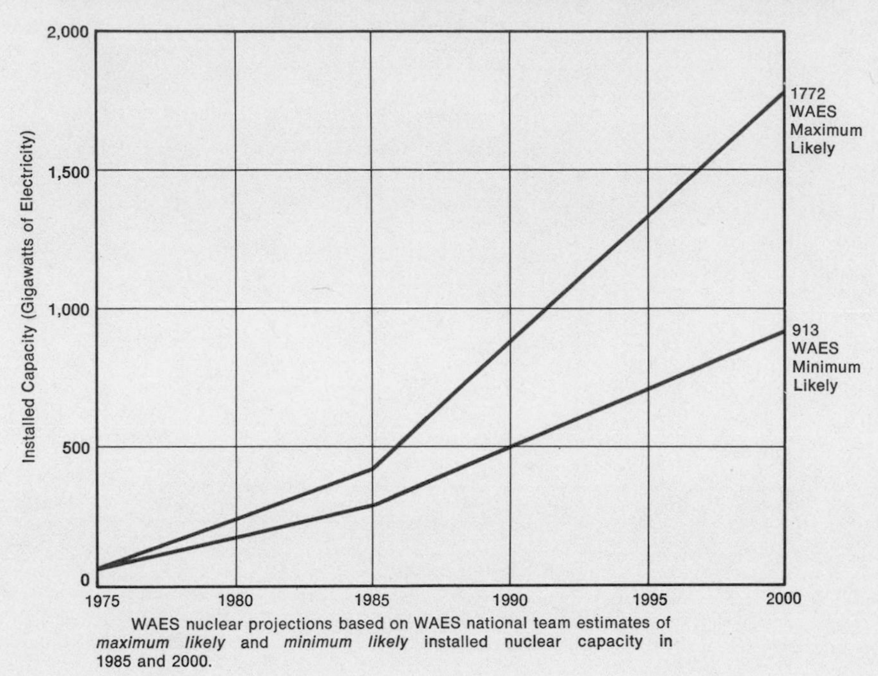

WAES nuclear projections based on WAES national team estimates of *maximum likely* and *minimum likely* installed nuclear capacity in 1985 and 2000.

costs. Because fuel cost is a small fraction of the cost of the electricity produced in nuclear power plants, increases in uranium costs have much less effect than increases in fuel costs in plants burning fossil fuels.

Another potential fuel, with reserves of approximately the same order of magnitude as uranium, is thorium. When thorium is placed in a region of neutron flux, e.g., within or surrounding a nuclear reactor core, thorium nuclei (Th-232) absorb a neutron and are transmuted into uranium (U-233) which is a fissile material that behaves like U-235 and Pu-239 in reactors. Thorium conversion to U-233 could become important in various types of thermal converter reactors, including heavy-water moderated thermal reactors and high-temperature reactors, if major research and development were undertaken.

Figure 6-5 shows estimated annual uranium requirements and Figure 6-6 shows cumulative uranium requirements to the year 2000 to meet the needs of the WAES maximum and minimum likely projections. Estimates of planned and projected uranium production capacities are based on the OECD/IAEA Report and were made before

Figure 6-5 Projected World Annual Uranium Requirements in WOCA

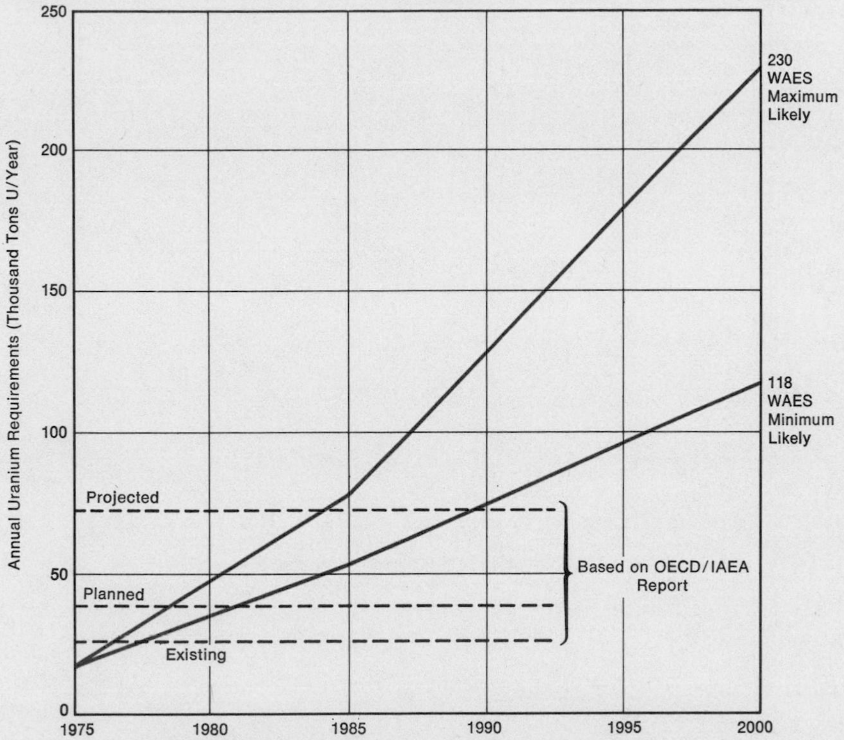

Uranium requirements depend on mix of reactor types, enrichment tails assay, load factor in operation, initial core, replacement loadings, plutonium recycle, delays in fuel cycle, etc. Required amounts shown here are computed on basis used in OECD/IAEA Report without plutonium recycle.

the uranium price rises of the past two years. A large amount of additional production capacity must be added to avoid uranium shortages in the early 1980's. From Figure 6-6, "Reasonably Assured Reserves" appear to be sufficient to meet uranium needs of the WAES maximum case until the early 1990's. Therefore, to the extent that "Estimated Additional Resources" are discovered and developed, uranium shortages would not appear until after the year 2000. The WAES low case could be met from Reasonably Assured Reserves until the year 2000. If electric power companies make long-term contracts covering uranium supply for 10-20 years of reactor operation, the total of RAR and EAR would be committed sooner at a lower level of installed nuclear capacity than in the WAES projections.

As in other parts of the nuclear industry, uncertainties about future demand discourage undertaking large exploration programs

Figure 6-6 Projected World Cumulative Uranium Requirements in WOCA

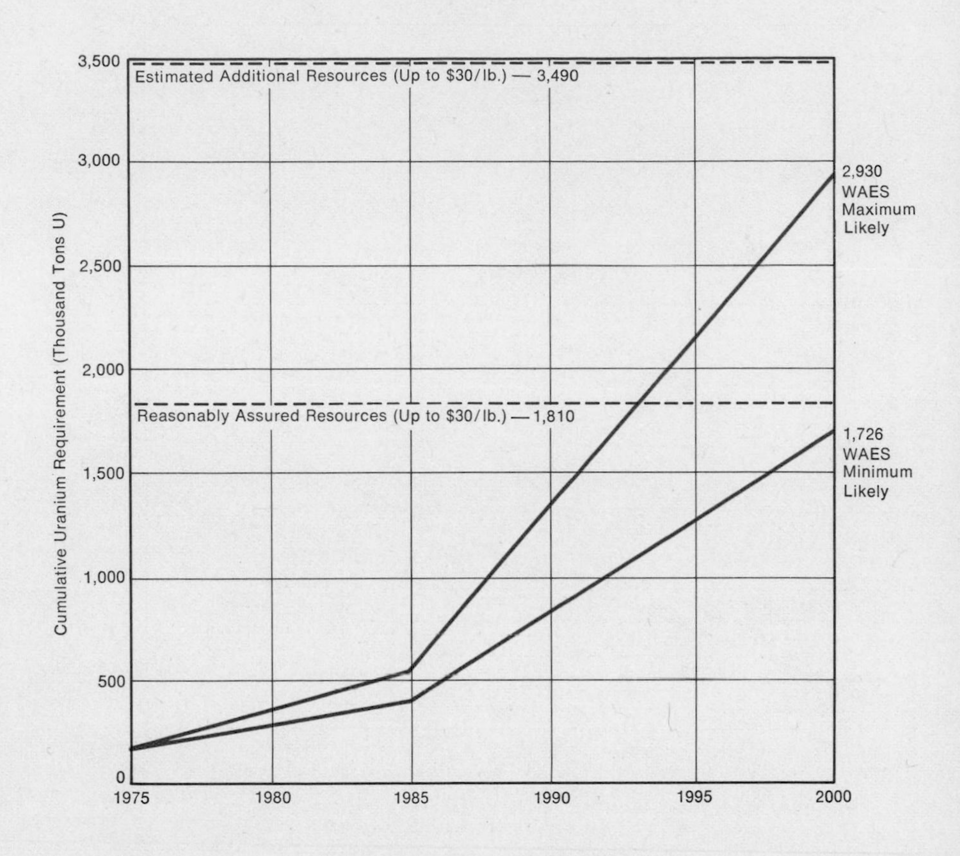

and the opening up of new mines, which may take 6-7 years to start producing and can cost as much as $100 million before the first uranium is shipped. Moreover, the uranium mining industry remembers the lean years of low demand and low prices of the 1960's. The Uranium Institute which has been recently formed may serve as a forum for producers and users of uranium and may lead to long-term forward commitments for uranium by the electric power industry. This would allow the uranium mining industry to make the necessary commitments and to raise the capital for the large additions to production capacity that are needed.

Availability of uranium will also depend upon the policies of uranium exporting countries. Canada, for example, earmarks for domestic use uranium reserves sufficient to fuel the nuclear reactors in Canada for 30 years. At this time it is not clear what conditions will be imposed on buyers of Australian uranium.

Enrichment Capacity

The unit of measure for enrichment capacity (a measure of the energy required for the enrichment process) is a Separative Work Unit (SWU). The enrichment process can be illustrated by Figure 6-7. Energy is applied to the natural uranium and the products of the enrichment process are enriched uranium and depleted uranium—called "tails." Separation or enrichment processes rely on the difference in mass between U-235 and U-238. In the centrifuge rotating at very high speed, the lighter U-235 is separated from the heavier U-238.

Figure 6-7 Schematic Representation of the Uranium Enrichment Process

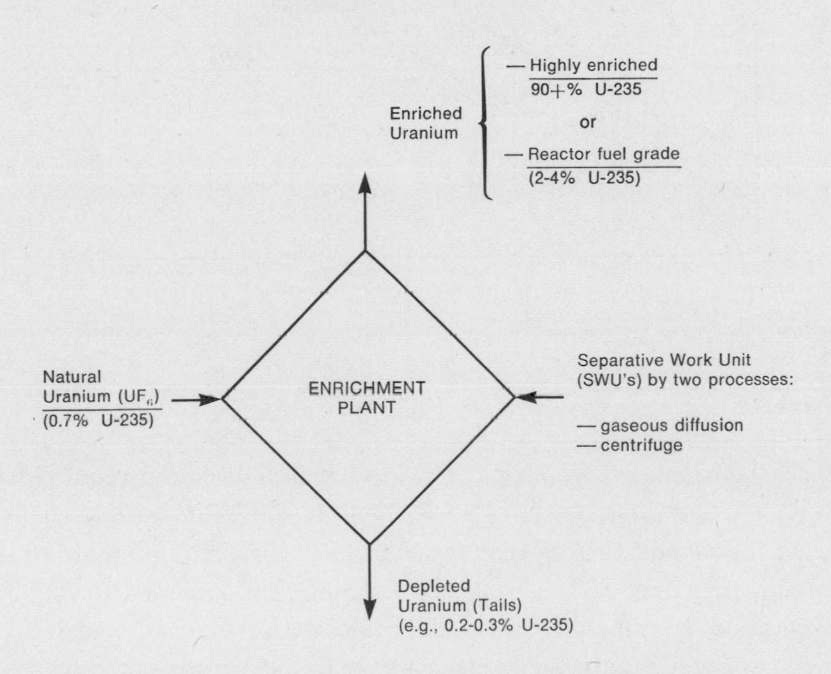

Estimated annual separative work requirements are shown in Figure 6-8. An increase in tails assay from 0.25% to 0.30% would reduce separative work requirements by about 10% and increase uranium requirements by about the same percentage. Conversely, reducing tails assay from 0.25% to 0.20% would increase separative

Figure 6-8 Projected Annual Separative Work Requirements in WOCA

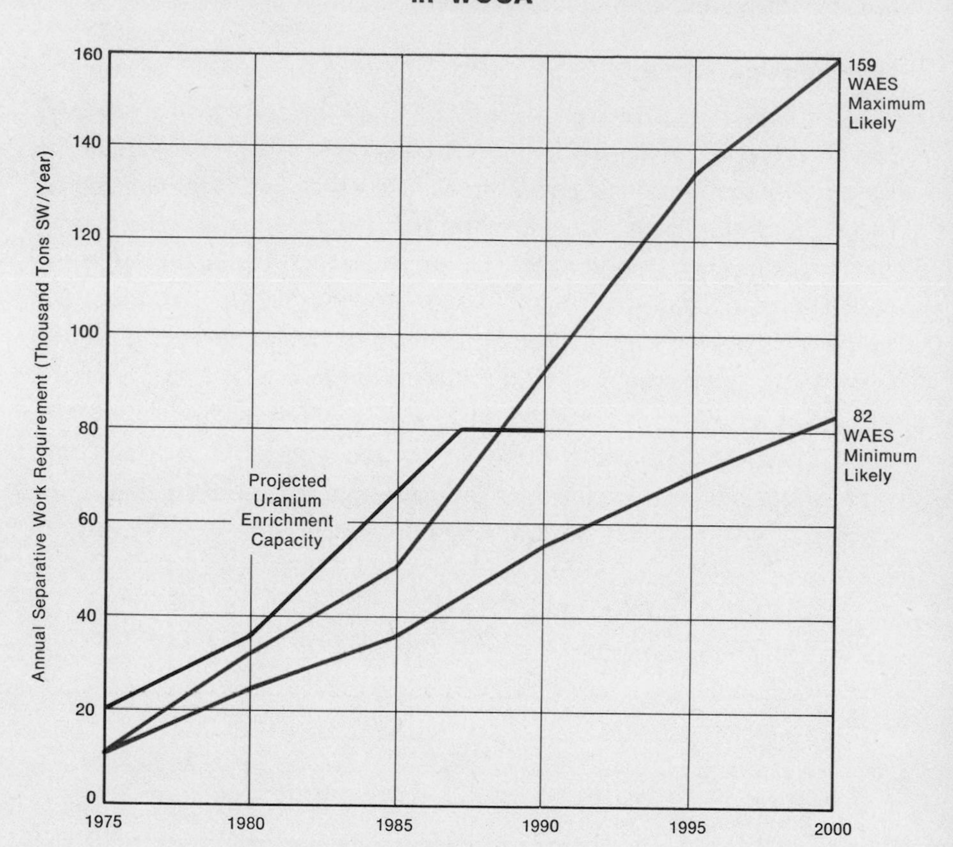

work requirements by about 10% and reduce uranium requirements by about the same percentage. Recycling plutonium recovered from used fuel would further reduce separative work requirements by approximately 10%. The projected uranium enrichment capacity was developed from the OECD/IAEA projections.

From Figure 6-8, it appears that if projected enrichment capacity is built without delay, the needs for enriched fuel for LWR's in the WAES maximum case could be met until the late 1980's, when additional capacity would be needed. Currently projected capacity could meet the WAES minimum case needs until the year 2000. Enrichment capacity must be sufficient a) to meet fuel needs for new plants, and b) to ensure a continuing supply of enriched fuel for existing nuclear plants.

Reprocessing Capacity

Table 6-3 shows projected additions to reprocessing capacity for LWR fuel as indicated in the OECD/IAEA Report, modified to reflect the announced cancellation of a 600 ton/year facility in the U.S.A. and delays on some other projects.

From such additions to existing capacity, we conclude that the uranium reprocessing capacity to recover U-235 and plutonium for LWR recycle or FBR initial fuel loadings will be insufficient until the late 1980's (assuming the WAES high nuclear case), thus requiring storage of used fuel in the interim. The WAES high nuclear case could still be feasible despite this reprocessing undercapacity, provided that interim storage facilities for used fuel are expanded which would allow storage of spent fuel for several years or more before reprocessing.

Table 6-3 Planned and Projected Additions to Projected Reprocessing Capacity for LWR Fuel (Tons Heavy Metal/Yr)*

Countries/ Projects	YEAR									
	1979	80	81	82	83	84	85	86	87	88
Europe	600	860					2600	4100	5600	6600
Japan	210								1710	→
U.S.										
AGNS	1500									→
NFS				750						→
Other								600	2100	4200
Total Capacity	2310	2570		3320			5060	7160	11660	14760

* From OECD/IAEA Report. Since the date of that report the NFS facility has been cancelled and other projects delayed.

Note: Assuming spent fuel output of 30 tons/yr per GW(e) of LWR capacity at 60% Load Factor, the annual output of LWR oxide fuel for reprocessing is shown below. The schedule of needed reprocessing capacity will depend upon the length of time spent fuel is in interim storage.

Year	1985*		2000	
	Max	Min	Max	Min
WAES Case	C	D	C-2	D-7
LWR Capacity GW(e)	371*	261*	1417**	728**
Spent Fuel/Tons U/Yr	11,130	7,830	42,510	21,840

* LWR = 90% of total
** LWR = 80% of total

Other Parts of the Fuel Cycle

While the basic technology for the LWR fuel cycle is well developed, some parts of the fuel cycle face complex problems that will have to be solved soon if the necessary supporting functions are to be adequate to support nuclear power on the scale and rate projected in the WAES assessments. Methods of disposing of radioactive wastes, in their ultimate repository, are still debated. If demonstrated systems for handling such wastes become a prerequisite for reactor licensing, WAES projections could be delayed by several years in countries with such policies.

Fast Breeder Reactors

Fast Breeder Reactors were projected by OECD/IAEA in their high case to amount to about 10% of total nuclear capacity by year 2000. We have reassessed the current position and probable future rate of development of the FBR in Europe. Our conclusion is that not more than 5% of the 533 GW(e) estimated for Western Europe in the WAES high nuclear case could be supplied by FBR's. By 1990, about 10 GW(e) of FBR's could be operating in France, Germany and Great Britain, with another 10-20 GW(e) possible by the year 2000. On such a basis, it is unlikely that power from FBR's could provide more than 5% of nuclear energy by year 2000. Estimated delays are due partly to schedules for construction and operation in order to gain experience with first-commissioned FBR's before additional units are ordered and built.

In the U.S.A. the situation is more uncertain, but recent official estimates* indicate that FBR's might begin commercial operation in the U.S.A. in 1995. To achieve and sustain the growth by year 2000 of a nuclear program the size of the WAES high nuclear case would require that the FBR become commercial around the turn of the century, or earlier. In addition, extra uranium supplies will be required to fuel the-then existing stock of thermal reactors during a transitional period until breeders begin to add to the fuel supply.

"Bred" plutonium from FBR's could add significantly to nuclear fuel supplies after year 2000, but schedules for such additions to fuel supply depend on the "doubling time" of commercial FBR's. The

---

* E.J. Hanrahan, "World Requirements and Supply of Uranium," presented at the Atomic Industrial Forum, 14 September 1976, Geneva, Switzerland.

"doubling time" is the time required for the FBR to "breed" double the amount of plutonium in the initial fuel loading. Depleted uranium stocks from the enrichment of fuel for the LWR's would be sufficient to provide U-238 for FBR's for a long time.

It is difficult to make a detailed technical assessment of the FBR at this time since very few reactors of this type are now actively operating in the world, and changes in designs will probably be made in the next generation of FBR's. Several countries are now planning to build FBR's because of their long-term fuel advantage. FBR's are needed to sustain a large scale nuclear power program after the end of this century.

Stages of Nuclear Choice

The debate on nuclear energy has made it very difficult for the public to understand that there are choices between accepting the entire nuclear system and foregoing entirely the benefits of nuclear energy.

Early expectations of the rate of growth of nuclear power implied such heavy demands on uranium that the projected program could only be met by introducing fast breeder reactors at a rapidly increasing rate by 1990. Such plans also required acceptable chemical processing of used fuel, plutonium handling on a large scale, and means for handling radioactive wastes. Conventional uranium reserves seemed unlikely to furnish enough fuel beyond the 1990's. Low-grade sources (below 100 ppm* U_3O_8) in shales (25-80 ppm); in granite (10-20 ppm); and in seawater (0.003 ppm) were not expected to furnish any substantial part of the projected needs. Given time and a major research and development program for exploration, mining and milling techniques for these low-grade uranium sources, it is possible that some of them could develop into significant sources in the medium to long term. But, they have not been regarded by WAES as commercial uranium sources in this century.

Three Stages of Choice

Lower estimates of projected nuclear build-up in most countries, with the accompanying reduction and stretchout of requirements

---

* Parts per million.

for uranium, does allow at this time a clearer view of the stages of choice. Some countries have already selected a course of action—others will have to choose. Table 6-4 below describes the three stages of nuclear choice.

Table 6-4 Nuclear Energy — Stages of Choice

Stage #1— Single Use of Uranium	*Stage #2— Reprocessing & Plutonium Recycle*	*Stage #3— The Fast Breeder*
— Single use of uranium	— Chemical reprocessing of used fuel to recover plutonium and uranium	— #2 plus fast breeder reactors
— Storage of used fuel without chemical reprocessing	— Handle plutonium	— Large-scale plutonium handling
— Defer decision on U-235 and plutonium recovery Disposal of liquid waste	— Disposal of radioactive liquid waste	(Breeders eventually multiply uranium fuel value 50 times)

Stage #1 — Single Use of Uranium

Three thermal reactor types have been developed and operated as full-scale commercial electric power plants: 1) the British Magnox and AGR; 2) the light water reactors (PWR and BWR); and 3) the heavy water reactor (CANDU). *Stage #1—Single Use of Uranium*—can be followed for all three reactor types by storing used fuel under water, where it cools thermally and the water acts as a shield against radiation. If at a later time it is decided to chemically reprocess the used fuel to recover plutonium and enriched uranium, such products will still be recoverable and chemical reprocessing will be somewhat simplified because of reduced thermal and radioactive levels of the used fuel. In any case, at some future time the used fuel must either be reprocessed or sent to permanent repository storage.

Stage #2 — Reprocessing and Plutonium Recycle

This stage requires acceptable systems for transport of used fuel; handling, transport and fabrication of plutonium into reactor fuel; decontamination of enriched uranium to levels acceptable in enrichment plants; and packaging and disposal of radioactive wastes. For LWR's, uranium recovery plus plutonium recycle might add fuel equivalent to roughly 20% more uranium.

Stage #3 — The Fast Breeder

Stage #3—The Fast Breeder—now seems essential if nuclear energy is to be sustained as a substantial source of world energy much beyond the end of this century. It implies acceptable resolution of all of the problems of Stage #2 and, in addition, the issues of safety of FBR's, which are different from thermal reactors. Implementing Stage #3 requires that decisions on Stage #2 be taken in time to provide the large quantities of plutonium needed for initial fuel loadings of FBR's.

Projections to the Year 2000

Choice in WAES Scenarios

The role of nuclear energy in the WAES scenarios is based on an assumption that most countries choose Stage #3—the Fast Breeder Reactor—although it is not expected that FBR's would add more than 5% to the nuclear energy supply by the year 2000. If Stage #3 is rejected or long-postponed for reasons relating to plutonium handling or waste disposal, Stage #2 might be similarly rejected or postponed, leaving Stage #1. The effects of such a sequence of events are difficult to estimate. However, unless uranium resources were to be greatly increased, or unless thorium became an accepted supplement to uranium, a long-continued election of Stage #1 would greatly reduce the future role of nuclear energy and lead to greater dependence on other energy resources.

Some Implications of Single Use

If single use of fuel were to become a standard mode of operation for nuclear power plants in some countries, uranium economy (the amount of useful energy obtained per ton of uranium) becomes of critical importance. Special consideration needs to be given to a comparison of LWR and HWR types in this respect. In such a comparison, the HWR appears at first glance to have a 4 to 1 advantage over the LWR because HWR fuel is a natural uranium (0.7% U-235).

In the enrichment process, for every five parts of uranium fed into the process, four parts remain as tails (depleted to 0.25% U-235) and one part comes out as LWR fuel (enriched to 3.0% U-235). On the other hand, in the HWR cycle all five parts of the uranium are used as fuel.

At the next step, however, the "burn-up" of uranium fuel in the reactor (measured in units of megawatt days per ton of fuel—MWd/T) appears to give the LWR a two to fourfold advantage over the HWR. If the burn-up is 15,000 MWd/T for LWR and 7,500 MWd/T for HWR, the HWR retains a net twofold advantage over the LWR in its economy in the use of uranium. On the other hand, if LWR burn-up is 30,000 MWd/T, the LWR has a fourfold advantage over the HWR as measured by "burn-up," and there is no net difference in uranium economy between the LWR and the HWR.* These are questions that deserve careful study if a single-use mode of operation is considered as a possible continuing policy in some countries.

Single use would also concentrate attention on the other potential reactor fuel—U-233 converted from thorium. Produced in thermal reactors, it might be brought forward enough to evaluate its utility and problems. If it can be used, thorium then could be added to the nuclear fuel resources. Thorium deposits are estimated to be of the same order of magnitude as high-grade uranium resources (RAR plus EAR), although exploration for thorium has been much less intensive than for uranium.

Implications of Uncertainty on the Nuclear Industry

This discussion of choices does not take into account the very large implications on the nuclear industry of switches among choices or of delays in decisions. Unless the environment for a vigorous development of the nuclear industry becomes more favorable soon, there is doubt that the levels of capacity projected in WAES scenarios can be met in time. The multibillion dollar scale ($0.4-1.0 billion for a 1 GW(e) plant), the long time interval (6-10 years) between start of a project and full-power operation, and the capital cost and lead times associated with other critical steps in the fuel cycle such as mines, enrichment plants, etc., make the dynamics of nuclear power

* Another method for estimating uranium utilization is by comparison of replacement fuel loadings—see IAEA Bulletin Volume 18, No. 5/6, page 8.

supply relatively slow to respond to changes in demand. The very high capital cost requires confidence among investors that such investments are sound and will pay out.

Nuclear Fusion

In the WAES time frame to the year 2000, it is not expected that power from a fusion reactor will contribute to global energy supply. It is uncertain whether a controlled nuclear fusion reaction of sufficient duration to be relevant as a power source will be demonstrated in the laboratory before the year 2000. In theory, controlled nuclear fusion of deuterium and tritium holds a promise of furnishing an almost limitless amount of energy. The engineering problems are formidable in using a heat source at 50 million degrees Celsius to make steam to drive turbines. The intense neutron flux in the fusion reaction makes everything nearby intensely radioactive. If a large and growing R&D effort is continued for several decades, fusion power might be added to the global energy supply in the 21st century.*

The Task of WAES

The reader will now appreciate some of the difficulties in estimating with confidence the contribution that nuclear energy might make to the world's primary energy needs in the year 2000. A vigorous program of building nuclear power plants, sustained over the next 25 years, could lead to nuclear energy contributing as much as 21% of primary energy in the year 2000. This would be equal to the energy contained in total WOCA oil consumption in 1975. However, unless the many obstacles now facing such growth are quickly and favorably resolved, the nuclear energy contribution will be less—perhaps much less. Differences between countries may become much greater than in the past as the people in each country decide on the risks and benefits of nuclear energy. Our assessment of the amount of primary energy that *could* be obtained from nuclear energy should also make clear the size of any shortfall. This shortfall would have to be met by other fuels if projected primary energy demand in the WAES cases for the year 2000 is to be met.

---

* An assessment of the problems and promise of fusion may be found in the TECHNOLOGY REVIEW, December 1976, The Prospect of Fusion by David J. Rose and Michael Feirtag.

CHAPTER 7

OTHER FOSSIL FUELS AND RENEWABLES

Other Fossil Fuels: Oil Sands; Heavy Oil; and Oil Shale — Renewable Energy Sources: Hydroelectricity; Geothermal; and Solar

Energy sources other than oil, natural gas, coal and nuclear could make a growing contribution to energy supply before 2000. The supply sources described in this chapter include other fossil fuels—oil shale, oil sands, and heavy oil; and renewable energy sources—hydroelectricity, geothermal and solar in its several forms.

Known deposits of oil sands, heavy oil and oil shale are much larger than the world's proven reserves of conventional oil. Such deposits represent a potential means for supply of petroleum long after conventional oil and gas fields are exhausted. To date, they have received very little attention compared with conventional oil, which has been abundant and inexpensive.

Oil Sands

In Canada, oil or tar sands lie in beds 50 to 100 feet thick under an area of 12,000 square miles near the MacKenzie River in northern Alberta. Both in situ and surface mining methods can be used to extract the estimated 300 billion barrels of oil considered recoverable. The amount minable from the surface depends on acceptable oil sand to overburden ratios.

At least 90% (some 270 billion barrels) of the oil sands lie too deep for surface mining and must be recovered by in situ methods. Active development of in situ systems, including pilot plant operations, is currently taking place. Capital costs for in situ systems might be substantially lower than for surface mining and the environmental problems of land reclamation involving very large quantities of sand would be avoided. Water flooding, thermal means and dilutents are being investigated as possible in situ methods which separate oil from sands in place, allowing the oil to be pumped to the surface. Technically feasible methods for separating the very viscous oil (called bitumen) from sand particles may present formidable problems for practical field operation. Hot water or steam must not leak into the surrounding sand or clay nor chemicals escape since they can be recovered and used again. The amount of energy contained in the injected heat or chemicals should be much less than the energy contained in the oil which is recovered. Otherwise there is little net energy gain.

Surface mining and processing of oil sands have been well demonstrated during several years of operation on a commercial scale at the Great Canadian Oil Sands (GCOS) where production is about 50,000 barrels per day (BD) of oil. A second 130,000 BD operation, scheduled to begin operation in 1978, will incorporate drag lines instead of bucket-wheel excavators as at GCOS. Processing of oil sands involves separation of oil from sand with hot water, partial refining to obtain several petroleum fractions, and desulfurization. Viscosity is reduced for ease of pumping at the low winter temperatures in the region. The oil is of good quality.

Problems of year-round operation in a hostile climate, excavating and transporting the sticky mass of oil sands and disposing of the very large quantities of sand tailings make such operations difficult and costly. Reserves accessible through surface mining are currently estimated to be about 30 billion barrels—comparable with North Sea oil reserves. If fully exploited, such oil sand reserves would support about 15 plants, each producing 200,000 BD, or a total of 3.0 MBD for 25 years.

These operations require an investment of about $20,000 per barrel of daily capacity (1975 Canadian dollars), resulting in a cost of $4.0 billion for a 200,000 BD plant. Thirty thousand man years are needed to build one plant. The cost of the oil produced is in the vicinity of present world oil prices, if royalties and taxes are excluded.

The rate of development of production from oil sands will depend upon trends in world oil prices, prospects for in situ recovery which might involve lower capital costs, and Canadian policy toward foreign investment. Given present and planned programs, it is unlikely that the oil sands production figures, which are technically possible, will be realized. Much more likely, based on present indications, is a Canadian production rate of some 800,000 barrels a day by year 2000.

Heavy Oil

Significant deposits of heavy oil have been found in Canada and Venezuela. Estimates of heavy oils in place are 167 billion barrels for the Canadian Cold Lake deposits, with a possible recovery of 15 to 30 billion barrels. Estimates for Lloydminster-type of crude oil are 10 to 15 billion barrels, with recovery of 2 to 4.5 billion barrels of oil. Recovery requires in situ systems somewhat similar to those needed for recovery of oil sands, or for tertiary recovery from conventional oil fields.

A complex accumulation of hydrocarbons, mainly heavy and extra heavy, was discovered in the south of Venezuela's Orinoco basin as a result of geophysical exploration carried out between 1925 and 1935. The boundaries of this area, called the Orinoco Oil Belt, have not yet been accurately established. Oil in place was preliminarily estimated to be 700 billion barrels in 1967. Information obtained after the first formal exploratory drilling program in 1973 demonstrated that the accumulations extended farther south than originally estimated. In addition, the thickness of the oil sands is greater than first estimated, and the crude's gravity is not low enough to indicate the presence of bitumens.

The Venezuelan Government is now considering the Orinoco Oil Belt's Integral Study and Evaluation Program; its objectives are to test and identify the most adequate methods for production and upgrading. The results of this program will be used to plan for the eventual and gradual commercial exploitation of this important oil-bearing accumulation.

Oil Shale

Oil shale is another potentially significant energy resource. The largest known reserves are in the U.S.A., with significant amounts in Brazil, Canada, Burma, the U.S.S.R. and China. Before the 1950's small oil shale industries existed in Sweden, the U.K. and a few other countries, but low-cost imported oil halted these industries.

Mining and in situ recovery techniques are two methods of extracting oil from oil shales. Underground mining by room and pillar methods in seams 100 feet thick, leaves about half of the shale in the pillars which support the roof. Substantial efforts are now being made to develop in situ production and a combination of the two methods is likely in the future. The oil in the shale, kerogen, is extracted from the shale by retorting the crushed shale at high temperatures and treating it with hydrogen to produce a semirefined fuel oil. Part of the kerogen decomposes into gas, which can be burned on site to furnish power or be converted into pipeline quality gas. Consolidating the spent shale from which the oil has been extracted and restoring the vegetation takes about 1.4 barrels of water per barrel of oil recovered. Of this, about a quarter of a barrel must be fresh water. This process may be largely avoided by in situ retorting in the mined cavern.

U.S.A. oil shale deposits are estimated to contain some 2,000 billion barrels of oil. However, only about 6% of the deposits are sufficiently accessible and concentrated (seams of thickness over 30 feet, yielding at least 30 gallons of oil per ton of rock) to be commercially interesting. Intermittent federal research and development and private efforts to develop oil from shale have been under way since 1948. There has been no commercial production to date, but a 10,000 barrel batch was produced for evaluation by the Department of Defense in 1975-1976. Table 7-1 shows current estimates of shale oil production costs in the U.S.A.

Table 7-1 Estimated Shale Oil Production Cost (1)

	Investment Cost $/Daily BBI	*Operating Cost $/BBI*	*Production* Cost $/BBI (15% DCF) 100% Equity*
Modified in situ	5,000- 7,000	3.50-5.00	8.00-11.00
Surface retorting	14,000-23,000	4.00-5.00	16.00-25.00

* The oil produced by these methods may be prerefined to a quality comparable with Arabian crudes for an additional $1 to $4 per barrel.

Economic, technical and environmental problems have hampered the development of shale. Inflation has increased the estimated cost of commercial plants beyond the range of usual industrial funding and uncertainty about future oil prices increases the investment risk. Environmental factors such as disposal of spent shale and water requirements present major obstacles to the development of a large oil shale industry in the U.S.A. In situ technologies which might diminish environmental impacts have not yet been demonstrated to be commercially feasible.

Federal subsidies for at least the early stages of oil shale development appear necessary to demonstrate the technology, economics, and environmental acceptability in a region where water is scarce and is largely committed for other uses.

The WAES scenario cases assume production of oil from shale of 2 MBDOE in WOCA in the rising energy price scenarios in 2000. Nearly all of the production is assumed to be in the U.S.A.

Hydroelectricity

There has been a steady increase in hydropower for electricity during the 20th century, and it is likely to continue to grow at a moderate rate to the year 2000. Most hydroelectric expansion is likely to be in the developing countries of the world. Together these countries have 44% of the world's hydraulic potential, but only 4% of this has been developed (2). Growth is somewhat constrained by the location of sites and by the long lead times involved with construction. We have therefore made conservative estimates of the future growth assuming that primary hydroelectric generation in developing countries will increase from about 1 MBDOE in 1972 to as much as 4.4 MBDOE by year 2000 (in the high-economic growth/low-nuclear case). This estimate could be higher if a number of developed countries, because of energy shortages or environmental reasons, choose to transfer some energy-intensive industrial activities to those developing countries which have abundant hydro resources, access to required raw materials and wish to develop such industries.

Growth of hydroelectricity in the developed world—which currently accounts for 80% of WOCA's hydroelectric capacity—will be less than in LDC's because most favorable dam sites are already developed. Growing environmental concerns may further discourage the development of all potential sites.

The WAES projections show a total world primary energy growth in hydroelectricity from 6 MBDOE in 1972 to as much as 12

MBDOE in the year 2000. Estimates of hydroelectricity growth by region for the WAES scenario cases are found in Chapter 8.

Geothermal

Geothermal energy depends on using thermal energy produced by various processes in the crust of the earth. Such energy is more concentrated in certain parts of the earth's crust, with the greatest amounts of potentially available energy in volcanic regions where tectonic shells meet. Stored energy may also be obtained in other areas where there are significant temperature concentrations caused, for example, by radioactive decay. Geothermal energy is not a renewable source outside the volcanic zones, and in these nonvolcanic areas, energy is drawn from a historically formed thermal store. A limitation of this energy source is that it must be used locally in the region where it is found.

Geothermal energy from dry steam is obtained by drilling in suitable areas. A more abundant source, but less efficient to use, is hot water (usually very saline), which is also obtained from wells. Geothermal fields, such as the Geysers (U.S.) and Larderello (Italy) require a combination of three geologic factors: 1) a natural underground source of water; 2) an impermeable layer above that traps the water and permits the formation of steam; and 3) a mass of hot rock near the natural water system. Only rarely do these three factors occur in the same area. Total installed global geothermal capacity was 1,400 MW(e) in 1974, less than the output of two nuclear power plants of 1 GW(e) each.

It has been estimated that exploitation of energy from hot dry rocks, without the impermeable caprock and the natural water system, could create a supply of geothermal energy several times greater than that of geothermal fields now exploited (3). Some experiments in the U.S.A. involve fracturing hot rocks (creating a large surface for heat exchange), injecting water into the lower part of the fractured area, and recovering steam or superheated water from the upper area to use to generate electricity. Fractures have been successfully created by drilling into the rock and pumping water under pressure into a field zone of the well. Such hydrofracturing of the igneous rock is a method from secondary recovery technology for petroleum (4). However, numerous problems remain. The fracture system may leak water, thereby preventing a build-up of pressure and recovery of heat and drilling in hot igneous rocks is difficult and costly. These and other potential problems must be solved in order to make hot-rock exploitation economically viable. By the year 2000, better answers to these technological problems will be known.

If research, development and demonstration enable commercial success by 1990, electricity production from hot rocks by the year 2000 will still be in its initial stages because of the time involved in exploitation, drilling, and development of geothermal fields and the subsequent integration of the developed fields into electricity generation and distribution system. Geographic areas of greatest hot rock potential include those of conventional geothermal energy production (Italy, U.S., Mexico, New Zealand, U.S.S.R., Iceland, etc.) and other regions of recent volcanism.

Solar

Since the energy crisis of 1973 there has been a growing belief by many people that the sun must become a major source of energy. Solar energy can be collected and used in many ways. Usually, radiant energy from the sun is used *directly*—in thermal applications such as heating water and space heating. Solar energy can also be converted into electricity by photovoltaic systems, or by mirror systems to produce steam for power stations. There are also *indirect* uses of solar energy through the use of wind and waves and using temperature differences between surface and deep waters in the oceans.

Each year the world's solar energy income at ground level is about 500,000 billion barrels oil equivalent—about 1,000 times the energy of the known reserves of oil but efficient extraction of this abundant resource is difficult. Solar energy is a very dilute source, and it is available only at intermittent intervals. In addition, solar climates vary greatly. In tropic latitudes sunshine is abundant and demand for space heating is small, while in more temperate latitudes winter days are short and overcast skies reduce the solar radiation reaching the collector. The need is to overcome these obstacles by finding ways to reduce the collector area required to capture this dispersed energy and by developing efficient systems to store collected solar heat for use at night or during overcast periods.

WAES projections assume that solar will contribute 1 to 2 MBDOE for all industrialized countries in the year 2000. Most of this amount comes from solar space heating and hot water systems. Significant amounts of electricity generated from solar sources seem unlikely before 2000.

Our solar projections result largely from the WAES energy price assumptions. Higher prices than in our scenarios would probably bring about larger contributions from solar energy. WAES studies do not look beyond 2000—at which time solar technology may be expected to make a growing contribution to the world's primary energy needs.

The rate at which solar energy will become widely used is highly uncertain, for technical, economic and institutional reasons. Research and development have only recently begun, making it difficult to forecast how well and how fast modern technology will overcome these barriers. Questions of costs, materials requirements, reliability, and acceptability need to be answered.

The financial constraints to widespread use of solar energy involve the capital needed for technological development, and the generally high initial costs of the systems themselves. Life cycle costs, in which high initial costs are offset by minimal operating costs, could give solar systems a growing advantage over systems based on fuels with rising prices.

These limitations vary greatly among solar energy's many different forms. Likewise, the extent to which modern technology has the potential for overcoming these barriers varies considerably among the different types. The growth of each type of solar energy can be expected to exhibit special dynamics.

Despite such problems and complexities, the potential of solar energy is such that after the year 2000, solar can be expected to play an increasingly important role in the energy mix. A U.S. Energy Research and Development (ERDA) report, for instance, suggests that solar energy could contribute 25% of the total U.S. primary energy by 2020.

WAES has not prepared a thorough analysis of the global potential of solar energy comparable to our work for the major fossil fuels and nuclear. Limited time has forced us to restrict our study to WAES national teams' estimates of the contributions of solar energy in the WAES countries, yet we do believe that solar and other renewables should be researched, developed and demonstrated as a highest priority. Our limited treatment of this subject is not a reflection of the degree of importance we assign to solar and other renewable technologies. Indeed, we consider such energy forms to be of great importance in the long-term energy future.

Solar Uses for Hot Water and Space Heating and Cooling

In the climates of North America, Europe, and Japan enough solar energy falls on the roof of the average house to supply all of its energy needs. Capturing and converting portions of this energy for useful purposes appears economically feasible over much of this region, at least for certain applications.

The best-known application of solar energy is for hot water heating. Improved technology, lower manufacturing costs, and strong marketing efforts—enhanced by public policy—may lead to substantial entry of solar hot water systems into markets in areas where sunshine is strong during much of the year.

Systems for the solar heating and cooling of buildings will probably develop more slowly than solar hot water systems. For efficient use of solar heating, buildings should be properly oriented, well insulated, and equipped with supplementary heating and energy storage. Currently, small production levels for solar units usually result in high costs. When produced in greater volume and with more experience, capital costs should be substantially reduced, and maintenance and reliability requirements will be better understood.

Programs for applying modern technology to the collection of solar energy and for storage systems which allow economical recovery and use are now being started. Such collectors may be *passive*, such as heat-absorbing walls or roofs of buildings, or *active*, as when a moving fluid collects the solar energy and transfers the heat to warm the building or stores the heat to meet heating demands at night or during heavily overcast days. An active system involves collectors which should be efficient, have long life with low maintenance, and be aesthetically acceptable because very large collector areas will be needed. Much progress is being made in developing collectors and other parts of a solar heating system including pumps, controls, and efficient means for storage.

The turnover time of a nation's housing stock is approximately 100 years. Thus, progress in the implementation of solar heating in the industrialized world may be slow if applications of solar technologies are confined to new construction. For example, to achieve a solar energy usage equivalent to 1 MBD in the U.S.A. by the year

2000 would require the building of solar hot water and heating systems in 2 out of every 3 homes built between 1980 and 2000. Such an effort would satisfy only about 2% of the total U.S. energy needs in 2000. A key problem, then, is how to devise more economical solar designs for retrofitting existing buildings and to provide incentives for such investments with high first cost and very low operating costs.

Rising prices of heating fuels would make solar systems (which need no fuel other than for the backup system) increasingly attractive. Indeed, the lifetime heating cost of a solar system may be much less than the lifetime cost of a conventional heating system requiring fuel. But most of the world's economic and financial systems do not take account of lifetime costs—total costs during a 20- to 30-year period. The building owner must make a large investment; consequently lowest first cost is typically favored. Moreover, the capital cost to provide electricity, gas or oil is now borne by the supplier—not the user. Revision of current financing and other arrangements may be necessary for solar heating and hot water systems to become widely adopted. Moreover the buyer must be able to look to organizations to supply solar systems which can reliably guarantee long life and low service cost, because it is on these merits that solar heating is attractive.

Electricity from the Sun

There are many possible technologies available for the conversion of solar energy to electricity. Schemes are under consideration for producing electricity both on a *decentralized* scale and on a *centralized* scale (central station power plants). It is unlikely that any of these schemes will provide large amounts of electricity by 2000.

Solar energy can be converted directly into electricity in solar cells made from special materials such as silicon crystals. Such cells have provided electricity in spacecraft. Much research and development is being done and breakthroughs on efficiencies and costs of cells in mass production appear possible. Again, storage systems and/or backup systems will be needed. Just as in the case of solar collector systems for heat, the cost of such systems for solar electricity will fall on the user. Since the capital cost of supplying electricity is now borne by the electric power company, new incentives may be needed to induce the user to make such an investment himself.

Two basic kinds of technology for central station conversion of solar energy to electricity are under consideration: thermal and photovoltaic (ground and space). Significant applications of these technologies appear to be highly doubtful before 2000.

One method involves concentrating sunlight on a boiler to superheat a fluid which then drives a turbine generator. Because of cost considerations, these designs may have to be relatively large to reduce the cost of power: 100 MW(e) and above.

Engineering feasibility studies indicated that "power tower" designs may be preferable. These use a field of individually guided mirrors (heliostats) to focus sunlight on a boiler at the top of a large tower. When generating electricity, these towers use only direct sunlight, and therefore are restricted to areas of low cloud cover. High capital costs, intermittent operation dependent on daylight, coupled with uncertainties in maintenance and lifetime, are major obstacles.

Studies are now being made of large central station photovoltaic systems. Implementation will have to await further breakthroughs in the cost of manufacturing solar cells.

Another proposal calls for satellites, placed in synchronous earth orbit, to convert solar energy into microwave energy to be transmitted to earth for conversion to electricity. This requires no new technologies, and avoids the problems of the earth's atmosphere and cloud cover. However, the satellites would have to be enormous—with solar panels in space more than 6 miles square—to produce 5,000 megawatts at the ground receiving station. The costs for transporting the required materials into orbit, and for building the system, are very large.

Solar Energy from the Seas

Three schemes proposed for the extraction of some form of "renewable" energy from the sea are briefly mentioned below. The capital requirements are large. The possibility of such sources making any significant contribution to meeting energy needs this century appears to be vary small and we have not included any energy from such sources in the WAES cases.

Ocean thermal systems would use the small temperature differences (about 40°F) between the sun-heated upper ocean layers and the colder, deeper waters to generate electricity. The temperature difference is largest in the tropics. While there appear to be economic possibilities for such schemes, the required systems are large and expensive, and must operate in a very unfriendly and corrosive environment.

Other systems harness the energy in ocean *waves.* Although the total amount of energy stored in waves is much smaller than that in some of the other forms of solar energy, it might provide modest amounts of energy in certain countries (like Great Britain) where much of the research in this field is currently being done.

Although the *tides* result from earth-moon, rather than earth-sun interactions, they are possible "renewable" sources for power generation. Two existing modern tidal power systems—one 240 MW(e) scheme at La Rance in France and a smaller one at Kislaya Guba in the U.S.S.R.—have demonstrated that tidal power can be economically harnessed. However, since there are only a few major estuaries in the world which offer prospects for moderately large tidal projects, power from tides is expected to be small and localized.

Bioconversion

Each year, 17 times more energy is stored in plant matter (by photosynthesis) than the world now consumes. The use of municipal, agricultural, and forestry wastes, growing plants for fuel, and other methods for extracting solar energy from organic matter constitute a potentially significant resource.

Growing plants for fuel is a novel, but potentially promising, approach, especially in developing countries. Tree plantations and floating kelp farms are variants being studied. Land-based energy plantations may compete with agriculture for limited arable land. Agricultural wastes in some cases may be sufficient in quantity and density to be used as a fuel, as bagasse from sugar cane has been used for many years.

Burning municipal wastes for energy or conversion into methane is a standard practice in many parts of Europe. This is beginning in the U.S.A., but it will probably require some form of tax credit for waste disposal to make it financially attractive.

Wind Power

Wind power is much less common now than it was 50 or 100 years ago, when it provided a significant portion of the energy consumed in rural areas. Yet there are many possible ways of extracting useful energy from the wind. Substantial research and development is now devoted to many modern windmill technologies (in blade dynamics and engine control)—with some emphasis on devices with capacities of at least 1 MW(e). New concepts are focusing on the use of wind concentrators, diffusers, and vortex generators which increase ambient wind velocity and so decrease the size of the turbines, the most expensive element in the system. Even with presently visualized cost reductions, the materials required suggest that electrical generators must either be large-scale or the price of electricity increase substantially, if winds are to become a significant source of power.

Even with large-scale, wind-driven generators, electricity output is small. It has been calculated that about 50,000 windmills with propeller diameters of 56 meters, and with average power of 500 W, would be needed to produce energy equivalent to 1 MBDOE.

Although the prospects for large-scale windmills are uncertain, local, small-scale uses could be important in many countries in the years ahead.

Solar: In Conclusion

Our brief summary of the potentials and challenges of several forms of solar energy is intended to highlight only some of the issues. We have not studied these topics in depth. We have treated some potentially critical long-term energy sources in cursory fashion; we have completely neglected other possible sources (such as wood, passive solar systems, and a wide variety of solar-electric schemes). Our purpose has been to place solar energy in context—globally and within our timeframe to 2000.

The institutional obstacles which now impede the rapid adoption of solar energy deserve more attention than they now receive. Research and development of solar systems is receiving increasing support and attention. It is now time that better institutional arrangements for widespread adoption be developed.

Other Sources: Conclusions

Our studies indicate that other fossil fuels—oil shale, oil sands and heavy oil—and renewable sources other than hydro will not make a major contribution to energy supplies before the year 2000. There is reason to believe that they could play an important role in the

21st century—if research, development and demonstration are pursued vigorously.

We hope that it has become evident that all of these potential energy sources must be investigated, evaluated and developed. The one certainty is that they all present hard choices, tradeoffs and potential opportunities.

References

1. *ERDA Weekly Announcements,* 24 December 1976.
2. *Energy and Petroleum in Non-OPEC Developing Countries, 1974-1980*, World Bank Staff Working Paper, No. 229.
3. Bernardo F. Grossling, "Geothermal Energy" Chapter 2, pp. 15-26 in *U.S. Energy Outlook*, National Petroleum Council, Washington, D.C. 1972.
4. Paul Kruger (Editor), "Geothermal Energy," Stanford University Press, 1975.

CHAPTER 8

CRITICAL PROBLEMS: ENERGY DEMANDS AND SUPPLIES MISMATCHED

Energy Balancing — WAES Supply-Demand Integrations — Overview: the Integrated Energy Picture, 1972-1985-2000; Oil Imports; Gas and Coal Imports and Exports; Regional Energy Balances; Fuel Switching for Energy Balancing; and Implications — Energy Balances and Imbalances in 1985 — Energy Balances and Imbalances in 2000 — A Case Example — Other Cases and Other Uncertainties

Energy Balancing

To many, the "energy crisis" is no more than an episode in history—a brief, unexpected and exceptional action by a group of Arabian oil producers in the autumn of 1973. It is not generally seen as symptomatic of any larger, or longer-term problems. Yet this book presents substantial evidence for the possibility of severe and continuing energy problems in the future—problems described in explicit terms in this chapter.

Such problems, we think, are not simply problems of scarcity. They are not simply problems of greed. They are problems of imbalance, problems of energy demand and supply mismatched under most desirable combinations of economic growth and energy prices.

In Chapters 2 through 7, we present the results of our work on future energy demands and supplies. This work was based on estimates of consumers' desires for energy and producers' potential to supply energy in 1985 and 2000 (resulting from varying assumptions about economic growth, energy prices and energy policies). The demand and supply estimates were made largely, although not entirely, separately from one another. Yet energy demand and supplies do not occur isolated from each other. Those who consume energy assume its availability; those who produce energy assume demand for their product. Energy demands and energy supplies must therefore be assessed together.

That is to say, energy demand and energy supply must be in balance. And, indeed, they always are. No more energy is produced (supplied) than can be used (demanded); no more energy can be used (demanded) than can be produced (supplied). Energy problems occur, however, when (for a given price expectation) consumers' *desires* and *preferences* for energy exceed producers' physical and economic *potential* to produce. This is a kind of imbalance. It could be a very uncomfortable kind of imbalance.

The interactions of energy demand and energy supply are complex. They involve consumer preferences for certain fuels based on the price, convenience, cleanliness, and reliability of supply of such fuels. They involve producers' decisions based on perceived demands, prices and costs—and on their willingness to make needed investments. They involve the potentials and limitations of the energy processing, refining, transport, conversion, and distribution system. And they involve the myriad of possible national political decisions that motivate and facilitate—or discourage and impede—a diversification of supply sources, or an emphasis on the production (or import) and use of certain fuels in preference to other fuels.

Comparison of future desired energy demands with potential energy supplies comprise what are called supply-demand integrations in WAES.

WAES Supply-Demand Integrations

WAES has done supply-demand integrations on both a national and on a global level. As described in Chapter 1, we have used two approaches for global supply-demand integrations. The first, the *un-*

constrained integration, assumes as given the fuel mix demand preferences of each nation indicated in the WAES national integrations.

These national integrations reveal the different national plans and expectations, under the scenario conditions, for demand and supply. They show, in some cases, how nations might respond to a global picture that is inconsistent—that is, to a global picture of prospective gaps caused by cumulative national expectations and desires. National integrations are done in a variety of ways; the methods and results are presented in detail in *Energy Supply-Demand Integrations to the Year 2000: Global and National Studies* (MIT Press, June 1977).

The unconstrained global integration sums the net desired imports and net potential exports of each fuel in each country—WAES countries and non-WAES regions. Figure 1-5 of Chapter 1 illustrates this procedure. The outcome is either a balance—i.e., export potentials of each fuel are sufficient to satisfy import expectations of that fuel—or an imbalance. If there is a balance, an allocation of fuels in world trade can be made, based upon national import desires and assumptions about which regions might logically consume the potential exports that could enter world trade. Simple allocation rules depending on established trade patterns may be appropriate here, as are the national consequences of such allocations. If demand and supply do not balance—if there is a shortfall of any fuel or of total energy—no world trade allocation is made. The size of the prospective global gap (or surplus) for each fuel is noted. Such gaps should be seen by nations as signals for actions—actions of the proper scale, timing and duration to avoid or minimize the gap.

The unconstrained integration, then, presents a global picture of supply-demand imbalances by aggregating national integrations.

The second global integration procedure is termed the *constrained* integration, since it takes explicit account of constraints in the energy system, including those of supply, through the use of a range of demand preferences for each fuel in each region and market. The procedure allocates fuels to regions within these preferences, taking account of supply constraints and costs, of the capital and operating cost of any expansion of energy production, processing, transportation, distribution and use infrastructure, and of the amount and costs of existing infrastructure.

The procedure is shown in Figure 1-6 of Chapter 1. National

and non-WAES countries' end-use energy demands are combined by regions. These regional demands are compared with regional supply potentials, to calculate a particular energy balance, with its own fuel mix, costs and infrastructure requirements.

The constrained integrations take account of the stages and routes—through production, processing, refining, transport, conversion, etc.—taken by specific fuels to meet specified desires for energy. Figure 8-1 illustrates schematically these stages and routes, from "Maximum Supply" (at the top of the chart) through various stages, with possible imports and/or exports, to meet "End-Use Demands" by consumers (at the bottom of the chart). The details of the flow between these supply and demand points in Figure 8-1 is simply an expanded picture of the circle labeled, "S/D Balance, Check, Allocation" in Figure 1-6. Such a flow exists for each fuel, with possible interconnections at the "processing and conversion" stage where one primary fuel may be converted to a very different processed fuel (e.g., primary coal converted to electricity). Integrations also reveal mismatches that may occur in any particular WAES case—prospective gaps where any fuel(s) may be in short supply relative to preferred demands for such fuel(s). This leads to regional fuel mixes which

Figure 8-1 Stages in Constrained Supply-Demand Integration

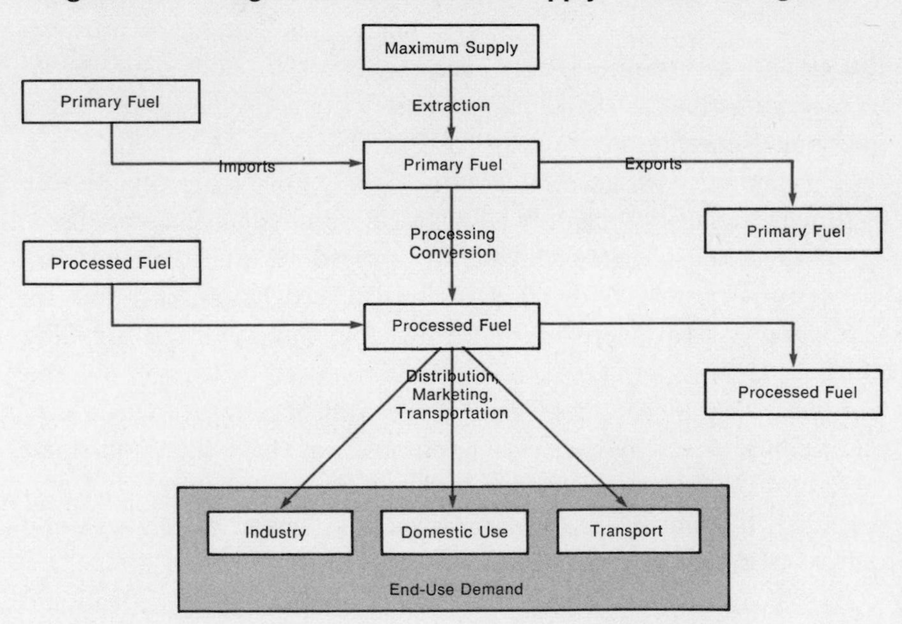

do not fall within market preferences in order to balance supply and demand.

In short, integrations give useful information, depending on whether there is a supply-demand balance or not. If there is a balance, then integrations reveal a wealth of detail about the energy system costs, efficiencies, allocations, and energy trade flows for a case. If no balance is possible, then the integrations merely signal the size of the imbalance, and the fuel or fuels found to be in deficiency or in excess. The following sections describe the results* of our energy supply-demand integrations. They represent the centerpiece of our work, the graphic picture of the fact that we live in an increasingly interdependent world.

Overview: The Integrated Energy Picture, 1972-1985-2000

Energy supply and demand, when taken together over the period 1972-2000, paint a disconcerting picture: growing shortages of oil. We illustrate this principal WAES finding for a high-demand, high-supply set of assumptions in Figure 8-2. Resource and production limitations begin to restrict oil supply in the period 1985-1990. Further increases in oil demand beyond 1990 must be satisfied from other fuels. The prospective oil shortfall must be filled or eliminated. Oil demand and supply must always balance.

Filling the prospective oil shortfall becomes increasingly difficult as we approach the end of the century. What begins as a minor discrepancy widens rapidly to a gaping deficit (some 20 MBD) by the year 2000. The picture persists under a variety of assumptions for the year 2000. Figure 8-3 illustrates the bands of estimates of both unconstrained** oil demand and potential oil supply over a range of assumptions. The problem does not go away within the range of the WAES economic growth, energy price, and national policy assump-

* There may be some minor inconsistencies between the results presented here and the figures in other chapters of this report. Any discrepancies will be the result of rounding, energy unit conversions, recent updating of estimates, or minor adjustments relevant to the supply-demand integration process. In no case will they result in material differences in the results of the analysis or impact the conclusions of the Workshop.

** "Unconstrained" oil demand is oil desired at the case assumptions, without regard to the actual quantity of oil imports that might be available.

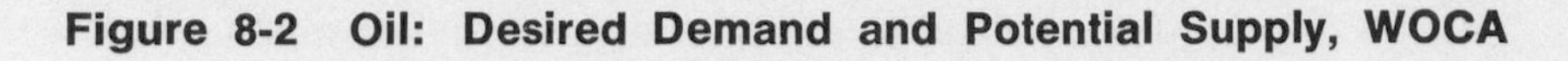

Figure 8-2 Oil: Desired Demand and Potential Supply, WOCA

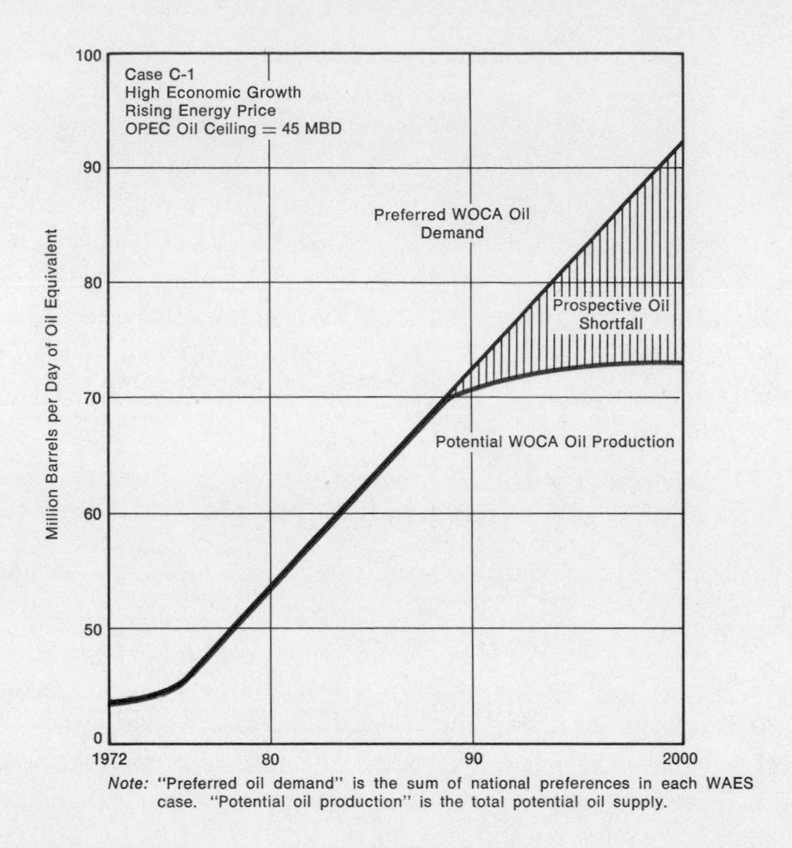

Note: "Preferred oil demand" is the sum of national preferences in each WAES case. "Potential oil production" is the total potential oil supply.

tions. Significant changes in our assumptions would, presumably, make the problem "go away." But it may only "go" to 2005 or 2010 —and then perhaps at substantial, even unacceptable, cost. We did not study such possibilities in depth. We encourage others to do so. Nonetheless, there are important lessons in the work we have done, and present in this report.

Oil Imports

The world oil problem, in simple terms, is the difference between desired oil imports and potential oil exports. Table 8-1 is a summary of WAES oil imports and exports estimated for the year 2000. It is based on individual national unconstrained supply-demand integrations from WAES national teams and similar assessments for non-WAES countries and regions (including OPEC). The WAES national unconstrained supply-demand integrations take account of con-

Figure 8-3 Oil: Ranges of Demand and Supply

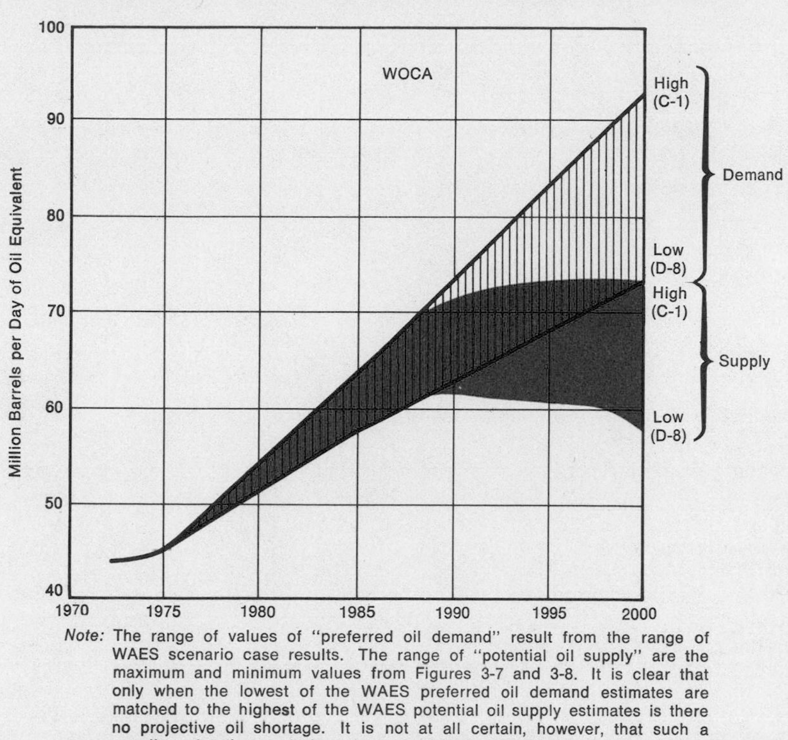

Note: The range of values of "preferred oil demand" result from the range of WAES scenario case results. The range of "potential oil supply" are the maximum and minimum values from Figures 3-7 and 3-8. It is clear that only when the lowest of the WAES preferred oil demand estimates are matched to the highest of the WAES potential oil supply estimates is there no projective oil shortage. It is not at all certain, however, that such a coupling of estimates is based on consistent assumptions.

sumer preferences and the resistance of these preferences to change (i.e., they recognize the lead times involved in consumers changing from one energy source to another). Desired oil imports exceed, of course, potential oil exports; with our assumptions, the prospective oil shortage is between 15 and 20 MBD.

The shortage is equivalent to an oil flow greater than half of OPEC's maximum production to date (30.5 MBD in 1973). It is approximately equivalent to the *total* amount of energy used in Western Europe today.

In each of our scenario cases this oil shortfall is nearly 30% of total WOCA potential oil production in the year 2000.

Notice that in the low-economic growth/constant-price cases (D-7 and D-8), desired oil imports by Western Europe, Japan and non-OPEC Rest of WOCA countries are *less* than their desired oil imports in the high-growth/rising-price cases (C-1 and C-2). This is due to the lower demand levels in the low-growth D-7 and D-8

Table 8-1 Summary of Oil Balance in Year 2000

Economic Growth:	High	High	Low	Low
Energy Price (1985-2000):	Rising	Rising	Constant	Constant
Principal Replacement Fuel:	Coal	Nuclear	Coal	Nuclear
WAES Scenario Case:	*C-1*	*C-2*	*D-7*	*D-8*
Major Importer's Desired Imports		(all numbers are in MBD)		
North America*	10.4	10.7	15.8	15.8
Western Europe	16.5	16.4	13.2	12.5
Japan	15.2	14.4	8.2	7.9
Non-OPEC Rest of WOCA	11.2	9.5	9.6	9.0
International Bunkers**	5.4	5.4	4.5	4.5
Total Desired Imports	58.7	56.4	51.3	49.7
Major Exporters' Potential Exports				
OPEC***	38.7	37.2	35.2	34.5
Prospective (shortage) or surplus	(20.0)	(19.2)	(16.1)	(15.2)
as a percentage of total WOCA potential oil production	27%	26%	28%	26%

* Takes account of domestic production of oil shale and oil sands in addition to conventional oil production.
** International bunkers represent the oil used in international shipping.
*** OPEC potential exports equal OPEC potential production minus OPEC internal demand.

North America: Canada, U.S.A.

Western Europe:	*WAES-Europe*		plus	*non-WAES Europe*	
	Denmark	The Netherlands		Austria	Luxembourg
	Finland	Norway		Belgium	Portugal
	France	Sweden		Greece	Spain
	F.R.G.	U.K.		Iceland	Switzerland
	Italy			Ireland	

OPEC (Organization of Petroleum Exporting Countries): Algeria, Ecuador, Gabon, Indonesia, Iran, Iraq, Kuwait, Libya, Nigeria, Qatar, Saudi Arabia, United Arab Emirates (Abu Dhabi, Dubai and Sharjah), Venezuela.
Non-OPEC Rest of WOCA: all other countries outside Communist areas.
WOCA: World Outside Communist Areas.

cases than in the high-growth C-1 and C-2 cases. For North America, desired oil imports are *greater* in D-7 and D-8 than in C-1 and C-2. This is due to the higher indigenous oil production levels of C-1 and C-2 (the higher-price cases) in North America, offsetting the greater demand in these cases. Among the major importers of Table 8-1, North America is the only one that is also a significant producer, and thus the only one in which production level differences can have noticeable impacts, from case to case, on desired oil imports.

The desired oil imports of 1972, two 1985 cases, and two of the four year 2000 cases are shown in Figure 8-4 for 3 major consuming regions. There is a continuing high level of dependence on oil im-

Figure 8-4 Required Oil Imports, as Shares of Total Energy Demand

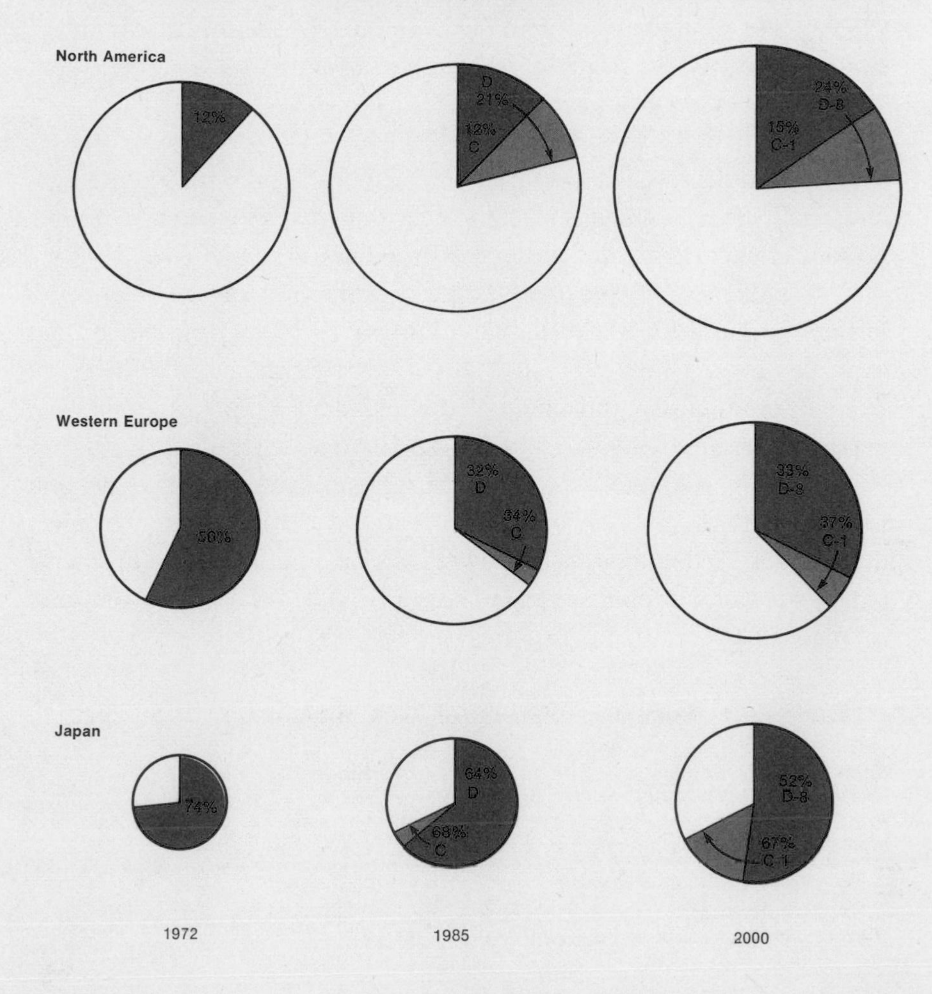

ports by Western Europe and Japan. In North America, imports grow as a fraction of the total in all cases owing to rising demand and dwindling U.S. and Canadian production. Yet even though oil imports represent only a relatively small proportion of the total primary energy requirement in North America by the year 2000, the import volume is growing. This growing North American drain on world oil trade has important consequences for other regions.

Let there be no mistaking our results: in spite of reduced overall demand growth, in spite of strong conservation measures, in spite of higher oil prices and (we feel) vigorous actions to bring on additional supplies, demands for oil continue to grow. The sum of national expectations for oil imports are very large in the year 2000—larger, as we have seen, than our estimates of the maximum potential oil exports. The result is prospective oil shortages.

Gas and Coal Imports and Exports

Nations also have import and export desires for natural gas and coal. These demands, which are calculated from unconstrained national integrations, are presented in Tables 8-2 and 8-3.

Table 8-2 shows the desired gas imports of the major importers, and the operational and planned projects from major gas exporters for our four year 2000 cases. The estimated imports result from assumptions in the national demand and supply studies. The exports from the U.S.S.R. and from OPEC and other Rest of WOCA countries are estimated based on maximum potential pipeline and LNG projects as of 1985. The increased gas trade required to meet desired gas demands is also shown. These requirements, in excess of operational and planned trade, range from 12 to 17% of the total

Table 8-2 Summary of Natural Gas Balance in Year 2000

Economic Growth:	High	High	Low	Low
Energy Price (1985-2000):	Rising	Rising	Constant	Constant
Principal Replacement Fuel:	Coal	Nuclear	Coal	Nuclear
WAES Scenario Case:	*C-1*	*C-2*	*D-7*	*D-8*
Major Importer's Desired Imports	(all numbers in MBDOE)			
North America	2.8	3.0	3.5	3.4
Western Europe	4.1	3.2	3.6	2.7
Japan	1.5	1.5	1.5	1.5
Total Desired Imports	8.4	7.7	8.6	7.6
*Operational and Planned Gas Trade**				
From U.S.S.R.	1.0	1.0	1.0	1.0
From OPEC and other Rest of WOCA	3.6	3.6	3.6	3.6
Total Potential Trade	4.6	4.6	4.6	4.6
Further trade required to satisfy desired imports	3.8	3.1	4.0	3.0

* Potential trade is based on those projects operational or under dliscussion at the end of 1976. These projects were all planned to be operational by 1985. Further trade is required to satisfy consumers' desired imports by 2000. In both 1985 and 2000 there could be a shortfall in gas supplies if projects are not developed at the required rate. Such shortfalls would be caused by lack of infrastructure rather than lack of resources.

Table 8-3 Summary of Coal Balance in Year 2000

Economic Growth:	High	High	Low	Low
Energy Price (1985-2000):	Rising	Rising	Constant	Constant
Principal Replacement Fuel:	Coal	Nuclear	Coal	Nuclear
WAES Scenario Case:	*C-1*	*C-2*	*D-7*	*D-8*
Major Importer's Desired Imports	(all numbers in MBDOE)			
Western Europe	4.2	2.1	2.0	1.1
Japan	2.3	2.3	1.7	1.3
Total Desired Imports	6.5	4.4	3.7	2.4
Major Exporters' Potential Exports				
North America	8.6	3.7	6.9	4.6
Australia, New Zealand, S. Africa	2.8	1.0	3.4	1.7
Developing countries	2.2	1.6	1.1	1.0
Total Potential Exports	13.6	6.3	11.4	7.3
Prospective (shortage) or surplus	7.1	1.9	7.7	4.9
as % of total WOCA desired coal demand	21%	6%	30%	23%

desired gas demand—or about 40 to 50% of total desired gas imports. The reserves of gas in OPEC are sufficient to meet major importers' desired imports, as well as provide for increasing domestic requirements. However, ways must be found to move the gas to markets. Alternatively, demands must be reduced through conservation or acceptance of substitute fuels.

Much more is said of the opportunities, constraints, and challenges of natural gas in Chapter 4.

Table 8-3 shows the prospective coal *surpluses* of our year 2000 cases. Coal is not, we find, a "preferred" fuel. The maximum potential for coal production exceeds desired coal demands by as much as 30% of total coal demand. However, nearly all coal demand estimates are based on known indigenous production estimates in each country. It is coal *imports* that are not preferred—so that potential global exports are greatly in excess of desired imports. Table 8-3 shows that the prospective coal surpluses are about the same size as total desired coal imports.

If demands for coal could be increased (as a substitute for oil and gas), the potentially available coal exports could be used to help close the oil gap. Strategies for the use, transport, and conversion of coal are obviously of critical importance. Further discussion of the issues surrounding the potential for world coal trade can be found in Chapter 5.

Regional Energy Balances

Our projections of energy supplies and demands—and imports and/or exports—for the major consuming and producing regions are summarized in the diagrams of Figure 8-5. These diagrams illustrate the relative import dependence of the consuming regions, and the potential exports from the producing regions, over the 1972 to 2000 period under the conditions of WAES Case C-1.

Figure 8-5a shows the growing rates of desired oil and gas imports to North America, at the same time that indigenous North American oil and gas production decline. As a result, oil and gas use decreases as a percentage of total energy demand, from 76% in 1972 to 48% by 2000 for this case (C-1). There are substantial increases in nuclear energy and in coal production and use, to meet growing demand and to replace (to a certain extent) oil and gas. North American coal production is sufficient, in this case, to allow

Figure 8-5 Regional Energy Supply-Demand Balances

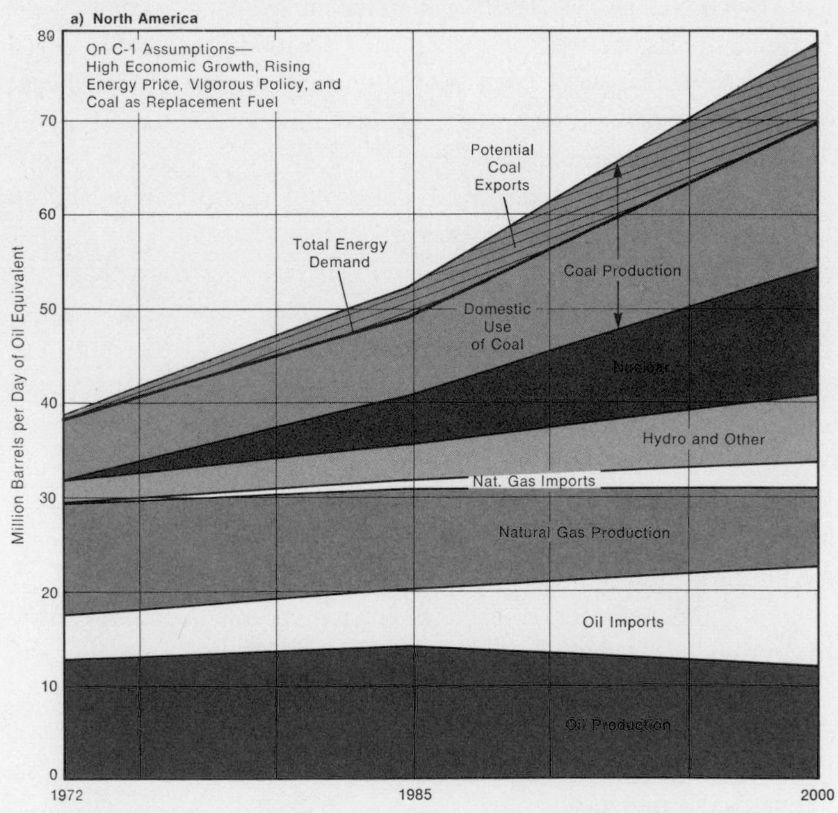

for growing amounts of *potential* coal exports—coal production in excess of internal demand that *could* be exported.

Western Europe would continue to increase its desired oil imports under the conditions of this case, from 12.0 MBD in 1972 to 16.5 MBD in 2000, as seen in Figure 8-5b. Both gas and coal imports increase during the period. Nuclear energy provides some 18% of Western European total energy demands in 2000 in this case.

Figure 8-5c shows that desired oil and gas imports continue to provide the major portion of Japan's energy needs to 2000. These imports account for some 73% of the increase in Japan's total desired energy demand by 2000. Most of the rest of the demand growth is met by nuclear (12%) and coal imports (9%).

As seen in Figure 8-5d, OPEC continues to be a major ex-

(Figure 8-5 cont'd)

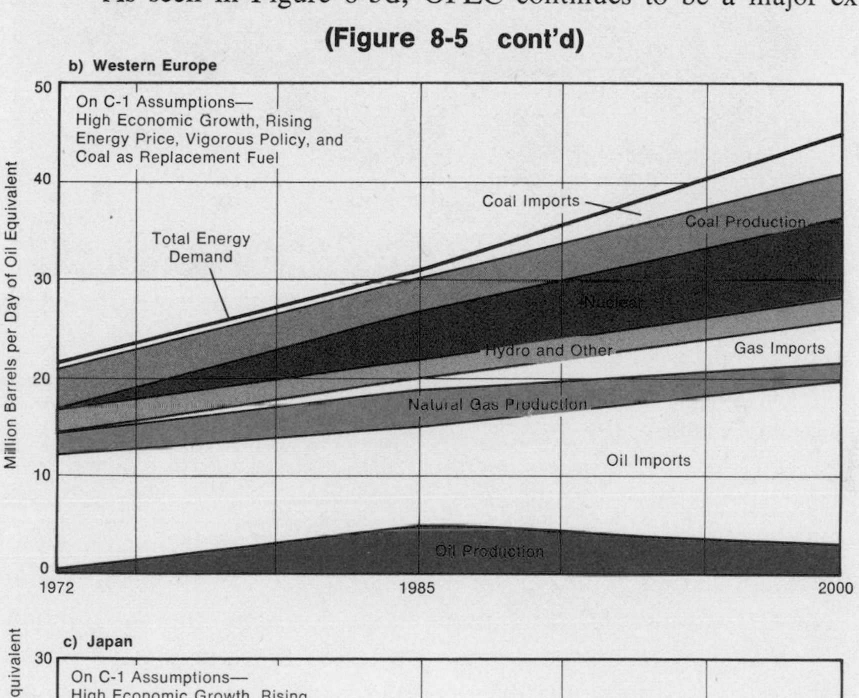

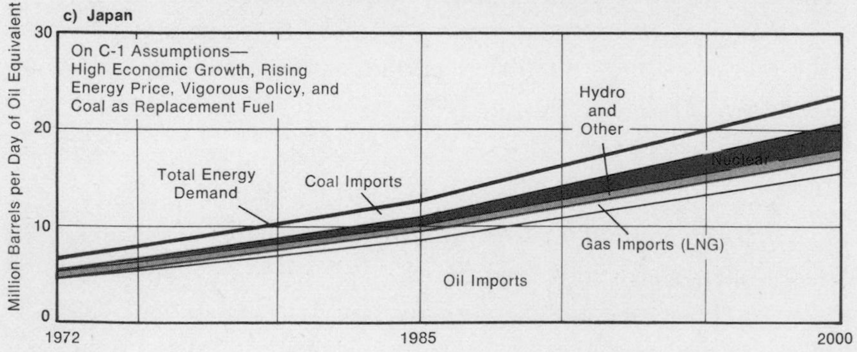

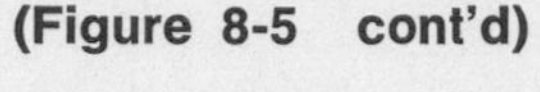

(Figure 8-5 cont'd)

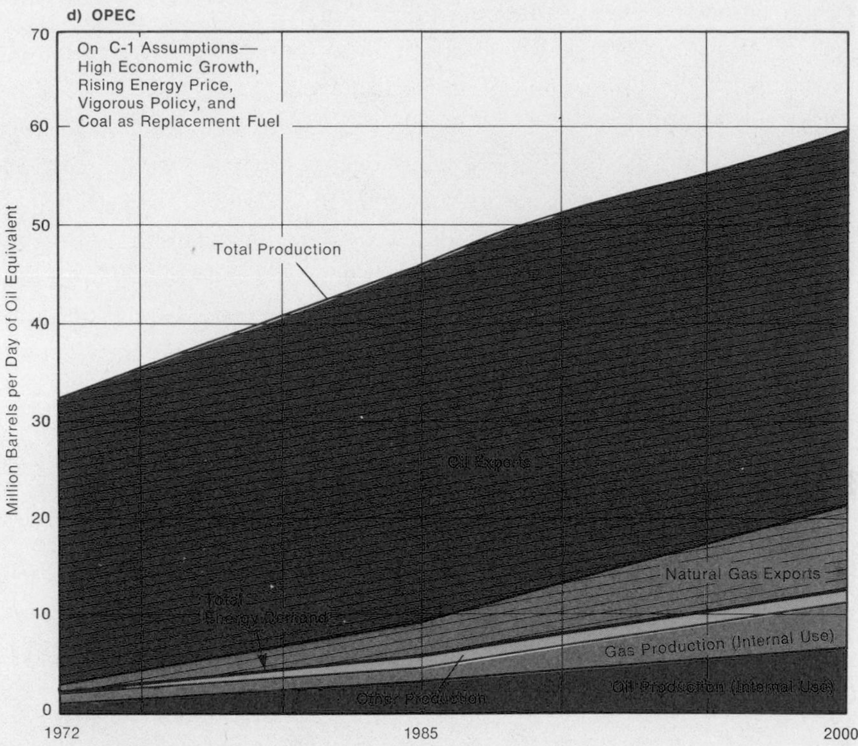

porter of oil, as well as a growing exporter of gas, for Case C-1 to the year 2000. Oil exports are nearly 39 MBD in 2000 in this case, compared to about 30 MBD in 1972. The potential production and export of oil and natural gas from OPEC are discussed more fully in Chapters 3 and 4.

Figure 8-5e illustrates the fuel mix projected for the non-OPEC developing countries for this case to 2000. Production and use of all fuels increase substantially with small amounts of coal and gas potentially available for export by 2000. In Appendix I we present a more complete discussion of the energy prospects of the developing countries to 2000.

Fuel Switching for Energy Balancing

The energy imbalances resulting from our supply-demand integration analyses cannot occur in the real world. Solutions will be found. But the cure could be more harmful than the ailment. Rapidly accelerating prices, rationing, economic stagnation, etc., all success-

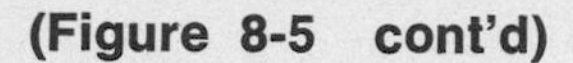

(Figure 8-5 cont'd)

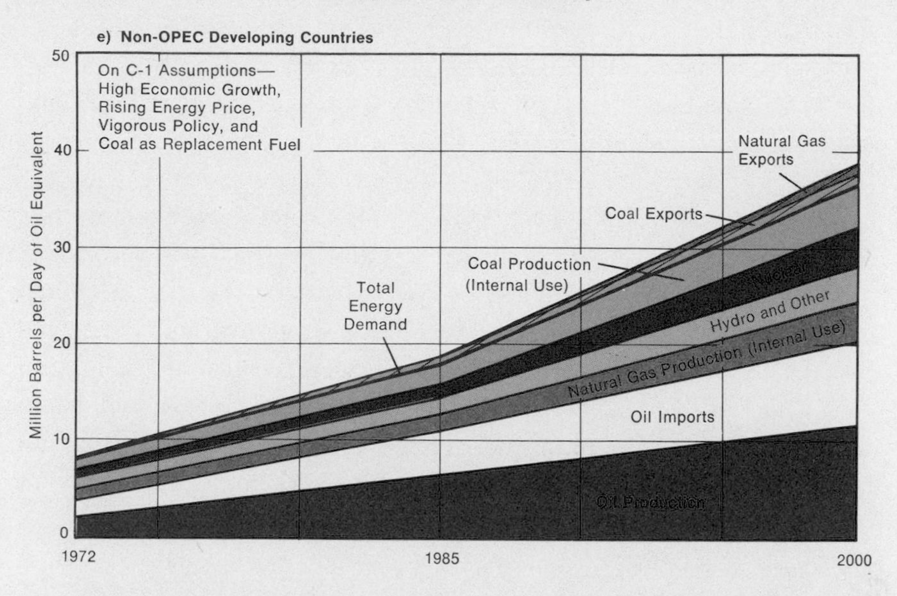

fully "reduce demand"; they balance energy supply and demand. Yet such dislocations and disruptions are very undesirable "solutions." They may be energy disasters, instead of energy strategies designed to smooth the transition from oil to other energy forms.

One broad approach to reducing energy imbalances is to ask: how might fuels be allocated so that, within reasonable constraints, the energy content of each fuel is put to maximum use? We tried to answer that question. We tried to squeeze every drop of energy out of each available fuel—minimizing the inevitable losses of energy content in processing, refining, and conversion—by using each fuel without conversion to other fuels, wherever possible. For example, the substantial heat loss associated with conversion of fossil fuel into electricity can be reduced by using fossil fuels in end-use markets *as* fuels—not in electric power plants—while using other options such as nuclear for electricity generation. (District heating from power plants is another example of a substantial efficiency improvement in this sense. Several national integrations incorporate such uses in their projections.)*

This efficient-energy-use allocation is part of the WAES constrained integration. Account is taken of fixed, nonsubstitutable uses

* See the national chapters in the Third WAES Technical Report.

of oil, such as transport and petrochemical feedstocks. For other uses, because of the severe constraints on energy availabilities, consumer preferences in the constrained cases often were not met. Nonetheless, the results are useful to show a potential mix of energy sources and uses that can contribute to solving the problem of prospective energy shortages.

The mix of fuels is important. That total energy supply and demand in WOCA differ only by a small percentage (between 5 and 8%) in our year 2000 cases is only a part of the problem. Small adjustments will not correct the real imbalance—the imbalance between the supply and demand of preferred fuels such as oil and natural gas. Our constrained integration procedure has made an effort to push substitutions for these fuels to the limits, in order to make up for the deficit in their supply. The resulting fuel mix is very different from what we are accustomed to today.

In our constrained integration, use of oil and natural gas in electricity generation in 2000 is marginal (about 10 to 15%) compared to a WOCA average of 35% in 1972; the balance is made up by a massive shift to nuclear and coal. Oil is essentially removed from use in the domestic sector and is replaced by electricity and by coal under various new technologies. Oil is also largely taken out of the industrial sector, except for use in feedstocks together with natural gas, and is replaced by coal to a large extent, and electricity to a small extent. Oil is used almost only in transport and for petrochemical feedstocks. Fuel substitution could not be pushed further even under the most optimistic technologies and vigorous assumptions about the speed of capital stock turnover—and yet supply does not quite meet demand.

Constrained integrations do not represent "predictions" of energy sources and uses. Rather, they represent the case of using all fuels in their most efficient way, within reasonable physical constraints, given the particular WAES case assumptions. They show what might be done, with concerted fuel-switching actions. They show the broad *directions* for action given by matching supplies and demands, to obtain an energy balance at the end of the century.

The results of the constrained integrations show that, given all of the assumptions and conditions (already described) of our analysis, the prospective energy gap can be reduced by contributions from the potential coal surplus, small amounts of synthetic oil from coal, modest exports from Communist areas, and processing losses saved by lowered electricity generation. Yet these measures, while moving in the right direction, do not close the gap. A small "unfilled demand"—a gap between 3 and 14 MBDOE in our year 2000 cases—remains.

Implications

The supply-demand integration analysis has 2 major implications:

1) All energy resources and conservation measures must be developed vigorously to meet total projected demand in the year 2000. Low levels of coal or nuclear development, or restrained conservation, would result in the failure of energy supplies to meet projected demand levels.
2) Incremental supplies in the 1990's depend on decisions taken in the next few years—a period likely to be one without severe energy imbalances. It is critically important to look *beyond* the next decade.

Within this global picture, individual countries will face a wide range of problems, of which the more significant are:

—Western Europe and Japan will continue to be largely dependent on imported energy.

—Within Western Europe there is a striking contrast between Norway, which shows energy self-sufficiency to the year 2000; the U.K., which shows self-sufficiency by 1985 but will be importing up to 40% of its energy needs by 2000; and the rest of Western Europe.

—The capital investment requirements for increasingly costly energy developments may require a higher proportion of nations' investments. In some of the developing countries, institutional support for these investments may become increasingly important. WAES has not studied this subject in detail. (Nor have we considered balance-of-payments questions in depth.)

—The criteria for assessing these capital investments during periods without severe energy stresses and for long lead time projects needs to be revised to reflect the lag between market price signals and the lead times required to implement needed energy supply and conservation programs.

Energy Balances and Imbalances in 1985

The critical energy problem for the world—the peak and decline of oil production in the face of rising oil demands—has already been elaborated in this volume and by others elsewhere. Of special importance is the "when" of the expected peak and decline. But within the range of WAES assumptions detailed already, the oil mountain's peak or plateau lies somewhere between 1981 and 2004. The year 1985 thus becomes of special interest for study. In addition, 1985 provides a detailed base year for projections to the year 2000.

Lessons from 1985

For 1985, energy supply-demand integrations reveal:

1) With a constant ($11.50)* oil price, and a range of real economic growth from 3.4% to 5.2% per year worldwide (WAES Cases C and D), sufficient supplies of energy are potentially available to meet total demands. In addition, demands for each fuel can be met with projected supplies. Thus, for these cases, energy preferences can be met with estimated supplies.

2) With an oil price rising to $17.25 by 1985 and economic growth from 3.4% to 5.2% per year (WAES Cases A and B), there appear to be significant potential *surpluses* of certain fuels over consumers' preferences for those fuels. The implication is that the high oil price ($17.25 by 1985) is probably higher than needed to bring on sufficient supplies. But actual surpluses would not exist in the real world. Either the price would be less than $17.25, demand would be greater than that at $17.25, or production would be cut back to match the demand.

3) With a falling oil price (to $7.66) and high economic growth (5.2%/year worldwide)—WAES Case E—there

---

* All oil or energy prices are given in constant 1975 U.S. dollars, per barrel of Arabian light crude oil, f.o.b. Persian Gulf.

are insufficient quantities of certain fuels available to meet consumers' preferences for those fuels. In addition, total potential energy falls short of total demand. Thus, for the high-demand, low-supply case in 1985, there are prospective shortfalls in certain fuels and in total energy. The case indicates that an oil price of $7.66, coupled with high economic growth, is too low a price to generate sufficient supplies to support the high demands resulting at that price.

Detailed results for the 5 WAES cases for 1985 follow.

Supply-Demand Integrations, 1985

Figure 8-6 illustrates the results of the 1985 WAES supply-demand integrations. Figure 8-6a shows desired demand and potential supply totals, for all energy, for the WAES 1985 cases. These demand totals represent the primary energy demand of a "preferred" fuel mix; they result from the sum of national unconstrained demands. The potential supply in each case is the energy equivalent of the maximum potential supplies of all fuels. Any difference between demand and supply in each case is the prospective shortage or surplus of *total energy*, when preferences are compared with potentials.

Notice, from Figure 8-6a, that only in Case E is supply insufficient to meet all demands. Unconstrained energy demand growth in this case gives 138 MBDOE in 1985 (4.3%/yr. growth), while maximum energy supplies are able to support only 116 MBDOE (2.9%/yr. growth). Total supplies would have to increase, or total demand decrease, by some 12 percent in order to balance this case. The combinations of low ($7.66) oil price and "restrained" national energy policy (tending to increase demand and tending not to induce increased supply), and high economic growth (spurring demand growth), is simply not a feasible combination by 1985.

Notice also that cases A and B show significant total energy surpluses, while cases C and D are nearly balanced in total energy in 1985. The higher oil price of cases A and B causes total supplies of all fuels to exceed desired demands by 11 and 25%, respectively.

The picture of gaps and surpluses becomes much clearer when we examine each fuel separately.

Figure 8-6b shows that nearly all of the energy deficit in case

Figure 8-6 Supply and Demand, WOCA, 1985

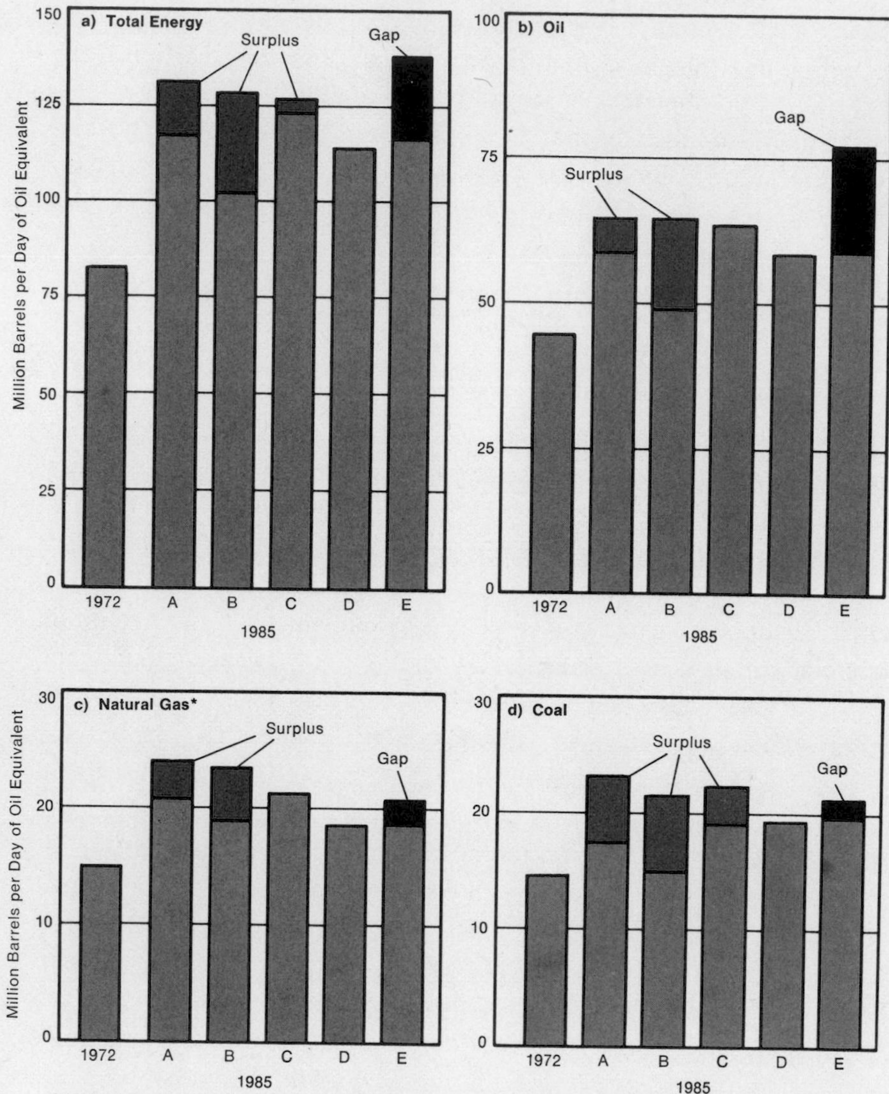

* Natural gas supply figures are *net* production; gross production figures would be slightly higher due to processing losses in gas transport.

E is the consequence of a prospective shortfall in *oil*.* Maximum oil availabilities in this case leave an oil shortfall of 18.2 MBD, or 24%, of a total unconstrained oil demand of 77.2 MBD. Notice also the

* The potential oil production figures for this case assume low gross additions to oil reserves (10 billion barrels per year) and an OPEC production ceiling

prospective surpluses of oil—surpluses of maximum *potential* production over desired demand—in the rising-price 1985 cases (A and B). These surpluses are large enough as a fraction of total available oil supply so that the cases should probably be considered as "infeasible." That is, the combination of variables that define cases A and B would not actually occur because oil production would not, in fact, exceed real demand but would adjust to it, through price, policy, and economic growth changes from the case assumptions.

The natural gas supply and demand balances are not unlike those for oil, as is apparent in Figure 8-6c. The falling-price case (E) projects a prospective gas shortfall of some 2.8 MBDOE. The constant-price cases (C and D) are balanced in gas supply and demand. The surpluses in the rising-price cases (A and B) are significant; but again, only enough gas would actually be produced to meet demand. The case assumptions would change, bringing demands and supplies of gas into balance.

Figure 8-6d shows the potential supplies and desired demands for coal in all 1985 cases. In most cases, potential surpluses are fairly small. In the low-price case (E), maximum coal production is slightly less (by about 7%) than desired demands for coal.

There are small potential surpluses of nuclear, hydro and other electricity-generating capability in all but the low-price, high-growth cases (E), in which all nonfossil electric capability is used.

A summary of the main observations resulting from our 1985 integrated cases can be found in the "Conclusions" section at the end of this chapter.

Energy Balances and Imbalances in 2000

Lessons from 2000

In nearly all of our cases for 2000, total desired energy demands exceed expected supplies. Even with the most efficient mix of fuels this is so; with "preferred" mixes, the discrepancy is larger. More importantly, there are significant oil gaps—prospective short-

---

of 40 MBD. It may be, of course, that OPEC would choose (if it had the technical and physical capacity to do so) to increase production levels enough to maintain constant revenues in this declining price case. The unlikely series of events leading to this might be the only way that the price would be forced down.

falls between potential (maximum) oil supply and preferred oil demand based on our assumptions. In short, the message from the year 2000 is one of prospective shortages.

But these results do not necessarily signal a bad future. The WAES year 2000 integrations signal a call for action—a call for concerted, consistent national policies and programs for dealing with the critical global energy problems. Our results are not predictions. They are intended as a guide to positive action, the basis for alternative energy strategies.

Even in one special case (discussed later in this chapter) where the energy balance is not severely stressed to the year 2000, the message to decision-makers in private and public sectors is still much the same: actions must be taken, and they must begin now. The transition from primary reliance on oil to other energy forms is inevitable. The transition may actually occur sooner or later than we project—under different conditions from those we assume in our cases. In any event, the available time must be used. It must be used to develop new technologies, to find new energy sources, to promote conservation—to adjust to the eventual transition. The prospective imbalance in year 2000 does not give a proper measure of the magnitude of problems that will occur in the 21st century if such measures are not taken.

Our findings for the year 2000 can be summarized as follows.

Supply-Demand Integrations, 2000

In Figure 8-7 we illustrate the spectrum of results for the year 2000 WAES cases. Figure 8-7a shows that total assumed energy-demand growth ranges from a low of 2.3% per year (Case D-7) to a high of 3.5% per year (Case C-2) for 1985-2000. In each of the four year 2000 cases, total potential supply of all energy is insufficient to meet total desired demands. These prospective shortages in total energy range from 5 to 8% of desired demand but it is not, as we said earlier, the deficits in total energy that are of most interest, or most meaning. Prospective shortages of *fuels* are the really important and revealing energy gaps.

The primary component of the energy deficits in the year 2000 cases is the prospective shortfall of oil, as Figure 8-7b indicates. For example, with rising energy price (to $17.25 by 2000) and high economic growth (4.0%/yr.) from 1985 to 2000 (Cases C-1 and

Figure 8-7 Supply and Demand, WOCA, 2000

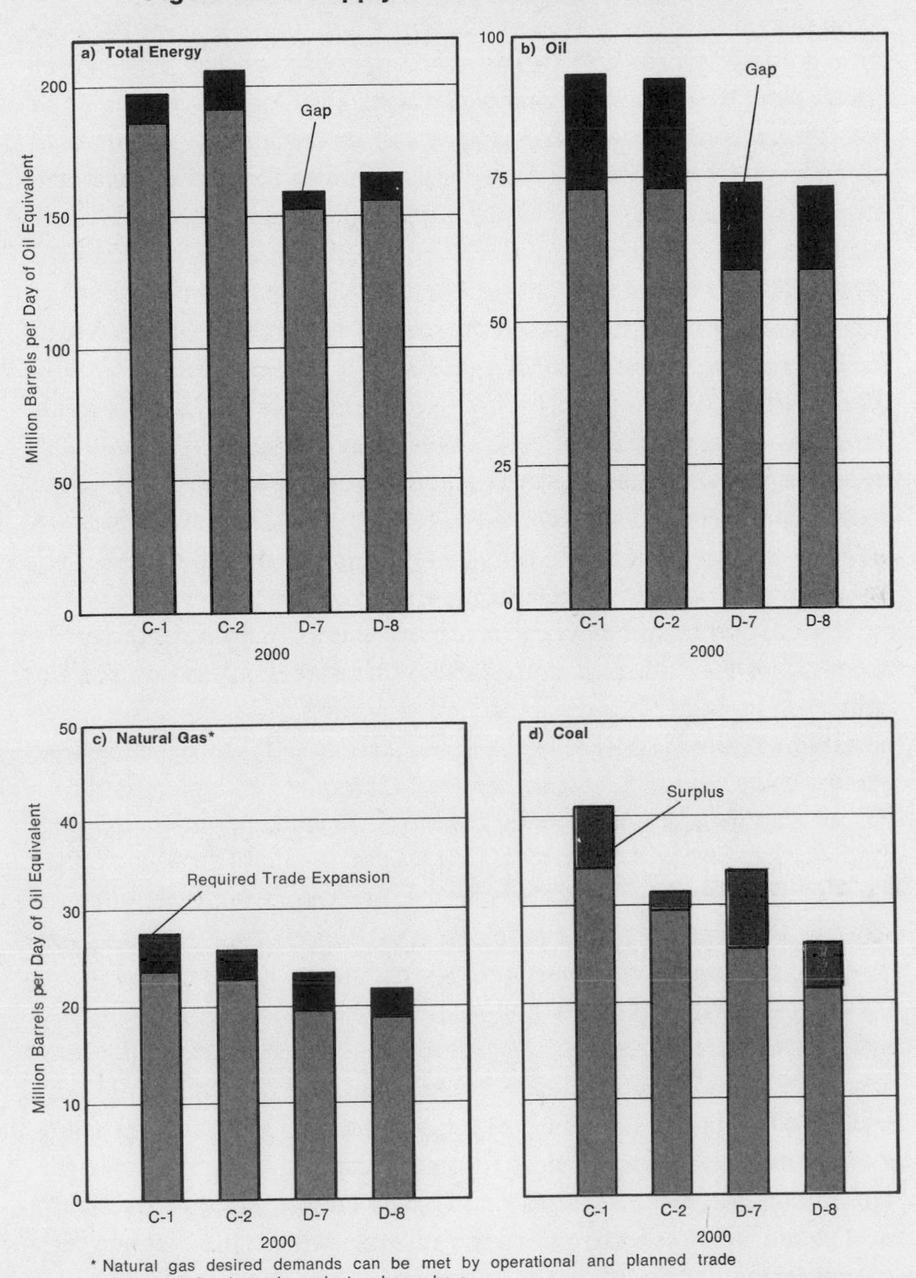

* Natural gas desired demands can be met by operational and planned trade plus expansion in trade projects, shown here.

C-2), potential oil supply in the world in the year 2000 falls short of the desired oil demand by some 20 MBD. In the WAES constant-energy-price ($11.50) and low-economic-growth (2.8%/yr.) cases to 2000 (Cases D-7 and D-8), the oil shortfall is some 15 MBD. These prospective shortages represent, in each case, about 30% of the total desired oil imports in WOCA. In other words, given the assumptions of the four WAES cases mentioned above, only about two-thirds of the aggregate desired global oil imports could be met by available oil exports from producers. World oil demand will probably overtake supplies sometime between 1981 to 2004 under any set of plausible assumptions.

The natural gas supply and demand balances are shown in Figure 8-7c. For all of these cases, some 3 to 4 MBDOE of expanded trade—trade in excess of operational and planned projects—would be needed in 2000 to meet desired gas demands. But global totals obscure the real point about natural gas. The gas supply-demand mismatch in the year 2000 is localized; the supply is not in the same geographical regions as the demand. Gas demands are unmet because of infrastructure limitations—limits on pipeline and LNG tankers—rather than because of resource limitations.

Oil and natural gas are "preferred" fuels. They burn relatively cleanly, and they are commonly used by many energy-consuming devices (autos, home heaters, power plants, etc.). Oil is easily transportable globally and locally. Yet supplies of oil are reaching and passing their zenith. Desired demands for oil of the magnitude of our projections will not be met. Demands for and supplies of gas are more complex (see Chapter 4). Gas gaps result from our projections, but these gaps *could* be closed. By the year 2000, the gas shortage is an infrastructure problem: gas is easily transported *within* a region, although not between regions.

The coal outlook is quite different. Coal is not a preferred fuel. It produces (substantial) pollutants when burned (which can be avoided at considerable expense), creates ashes to be disposed of, is not easily handled for transport, and would generally require many changes in existing user devices for its widespread adoption. Yet the world reserves of coal are enormous and largely untapped. Figure 8-7d shows the desired demands for and maximum potential supplies of coal for the four 2000 cases.

Under the conditions of the four WAES cases to 2000, maximum potential production of coal in the world exceeds desired coal demand by some 1.9 to 8.2 MBDOE (140 to 620 MT coal per year) in 2000. These prospective coal surpluses are between 10 to 50% of the prospective oil gap in each case. Coal thus is a potentially important replacement fuel for oil, on a global scale. (A more elaborate treatment of the potentials and challenges of coal can be found in Chapter 5).

The main conclusions that we draw from our integration analysis for 2000 can be found in the "Conclusions" section at the end of this chapter.

A Case Example

The characteristics of the global energy futures envisioned by WAES can be illustrated best by an example. Case C-1 (high economic growth, rising oil price, principal emphasis on coal) is selected for presentation here. It was not chosen because it is thought to be the most likely case or because it is the most typical case for year 2000. It is probably neither. General observations will be drawn from it wherever possible. Major differences between it and other 2000 cases will be noted wherever appropriate.

We examined a mix of actions to fill the "prospective oil gap" of our cases. Figure 8-8 shows, qualitatively, the results of that study. The major choices for WOCA are few. We could reduce the gap by a combination of:

—imports from Communist areas;

—coal (as coal) replacing oil in markets;

—synthetic crude oil made from coal; or

—losses reduced by fuel shifts away from electric.

It must be recognized that this constrained integration gap-filling is achieved by distortions of preferred fuel mixes, highest efficiency use of fuels, and least-cost allocations of fuels. The wrenching changes of these actions may be improbable. Yet low probability does not negate the validity and usefulness of this analysis, which uses a highly constrained linear programming routine—the Global Energy Mini-Model (GEMM). The analysis illustrates the potentials for actions, within the limits of the scenario, of *direction*—the direction of actions and programs that can contribute to solving the prospective oil shortfall. The indicated directions are much the same for other WAES year 2000 cases.

Figure 8-9 illustrates the international trade expansion re-

Figure 8-8 How Oil Supply-Demand Gap May Be Filled

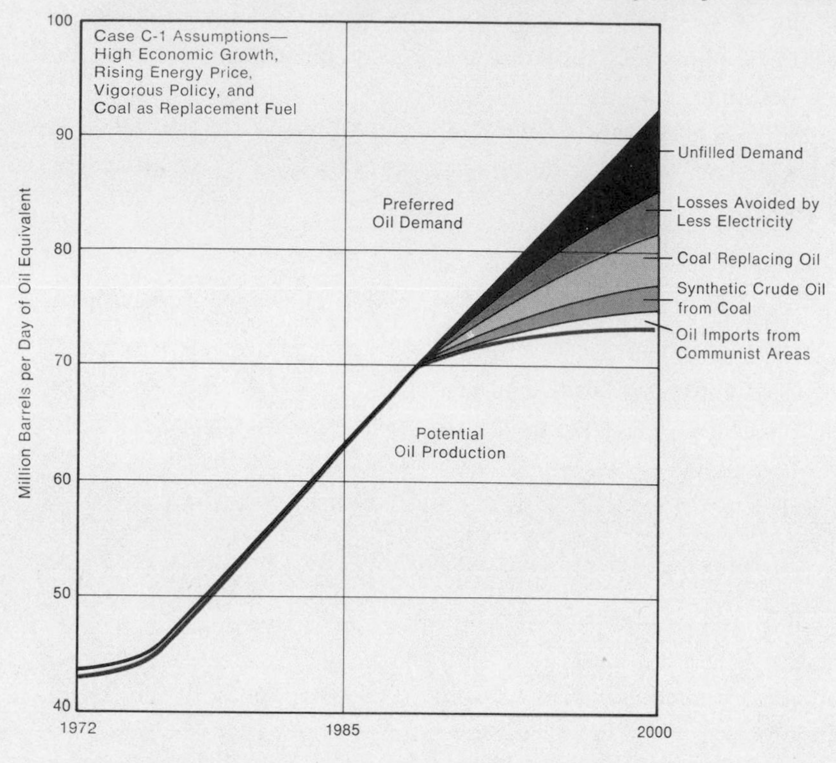

Figure 8-9 How Natural Gas Demands Could Be Met

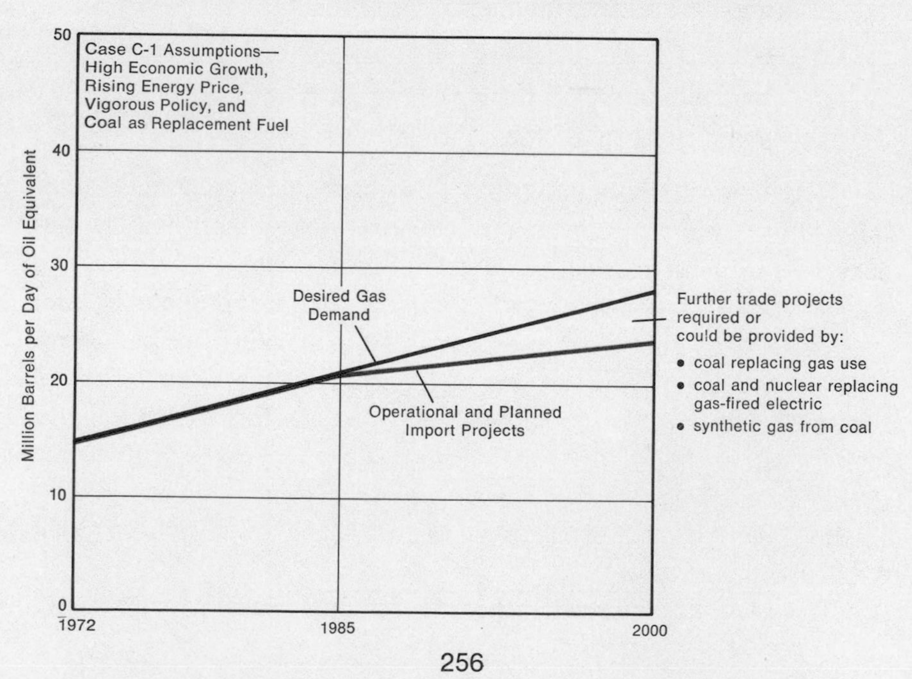

quired (about 4 MBDOE) if the major consuming countries' demands for gas are to be met. As discussed in Chapter 4, there are no resource limitations which prevent the export of such quantities, but political decisions may do so.

Any such limitation on supply availability would have to be filled by a shift in final user demand to other fuels or to synthetic gas. For example:

—coal (as coal) replacing gas use;

—coal and nuclear replacing gas-fired electricity generation;

—imports from Communist areas;

—synthetic gas from coal; or

—losses reduced by fuel shifts away from electric.

Once again, the utility of this analysis is its indications of the broad directions of effective changes. Replacement of gas-fired power plants, synthetic natural gas made from coal, and greater use of coal are general types of energy source and use patterns that can lead to solutions of a prospective gas shortfall.

The supply of coal is not in shortage in our cases. As Figure

Figure 8-10 How Coal Supply-Demand Surplus May Be Used

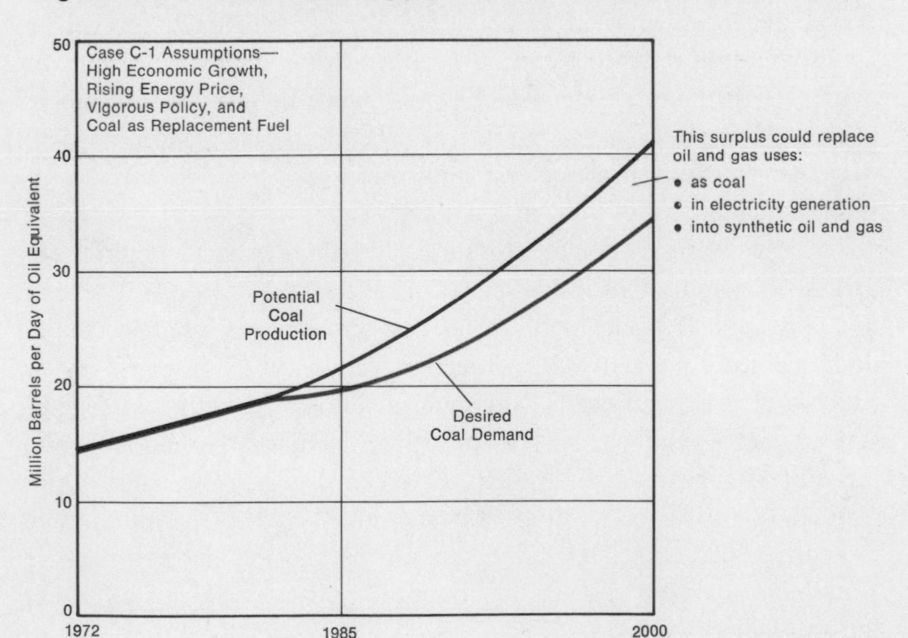

8-10 shows, the maximum potential production of coal exceeds consumers' preferences for that fuel. In case C-1, this potential coal surplus grows to some 7.1 MBDOE (540 MT per year) by the year 2000. Other WAES 2000 cases result in prospective coal surpluses of between 140 and 620 MT per year.

Obviously, coal can help greatly in filling the prospective gaps of oil and natural gas. But its widespread use greatly distorts consumers' preferred mixes. Coal could be used, as coal, in place of oil and gas in industry, in electric power generation, and perhaps in residential and commercial applications. Coal could be converted to synthetic oil or gas to replace conventional uses of these fuels.

Figure 8-11 illustrates the mix of fuels used in 2000 Case C-1 in three demand sectors: electricity generation, the domestic sector and the industrial sector. Two fuel mixes are shown. The first (the left-hand set of bars in each pair) shows the fuel mix obtained as the sum of individual national demands and preferences for fuels based on the scenario assumptions. This gives the demand-preferred mix. The second (the right-hand set of bars in each diagram) is obtained by attempting to reduce the overall demand for primary energy to match the available supply by shifting final consumer demand (using our linear programming model) between fuels, to use them in a more efficient way. This gives the supply-constrained mix. Fossil fuels in power plants are burned at about 35 percent efficiency; processing losses could be reduced by using fossil fuels directly (in heating, for example), in place of electricity. This approach gives a fuel mix of nuclear, hydro and "other" as sources for electricity generation, with domestic and industrial sectors using more coal and less oil and gas than the demand-preferred mix.

Notice in particular in Figure 8-11 that, in the industrial sector, global supply constraints result in a clear shift from oil to coal use; gas and electricity use are largely unchanged. Changes in the domestic sector, given these allocation rules, are a bit more severe. Electricity use is reduced—although it still, even with supply constraints, accounts for 22% of the domestic market in 2000—some 3.6 MBDOE higher than in 1972. Gas use is nearly the same as the demand-preferred level. Coal, however, enters in substantial amounts to replace oil uses. This result, which clearly and substantially violates consumer preferences, results directly from coal, not oil, being the fuel widely available.

Figure 8-11 Comparison of Unconstrained and Constrained Mixes in Demand Sectors

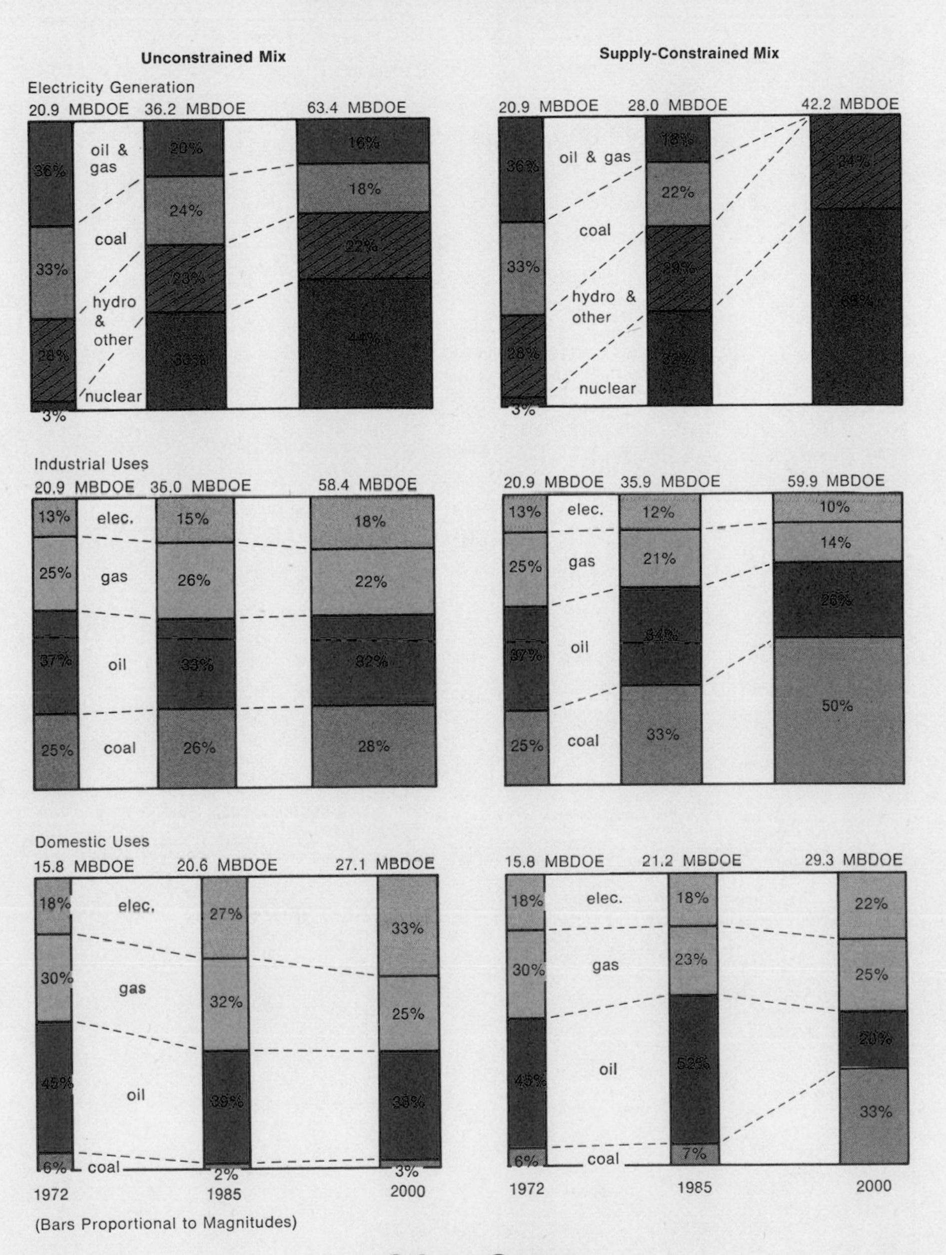

(Bars Proportional to Magnitudes)

Other Cases

The implications of the year 2000 scenarios just discussed make it clear that actions to reduce demand—particularly demand for oil and gas—and to develop supply alternatives should begin now;

"Actions Programs" to Close the Gap

The prospective shortages revealed by our analyses can be reduced or eliminated by what we have called energy "action programs." These programs represent greater efforts than we have already assumed to reduce energy demand, increase energy supply, and make substitutions of alternative fuels, such as coal and nuclear, for the most threatened fuels, such as oil and gas. These programs occur with prices above the case price, or with policies more vigorous than the case assumptions. The following arguments illustrate how the action programs could work.

The diagram to the right (1) illustrates, in the familiar price-quantity representation of microeconomic analysis, the prospective shortages of the WAES cases. Notice that at a sufficiently high price the sum of non-OPEC and OPEC supply increase to the point where the gap is closed ("stable price"). Further, unimaginably large price increases can lead (mostly through major demand reduction) to non-OPEC self-sufficiency.

Vigorous supply policy, *more* vigorous than assumed in our cases, can also potentially close the gap, as illustrated in Diagram (2) to the right. Such "super-vigorous" policies would induce supply-expansion action programs (at the case price, in this simplified representation) nearly sufficient to eliminate prospective shortages. The supply curves shift to the right, reducing the gap.

Super-vigorous conservation policies can also reduce the prospective oil shortages, as illustrated in Diagram (3) to the right. In this case, the more vigorous conservation action program drops the demand curve down, resulting in a lower demand at the same price. This reduces the gap.

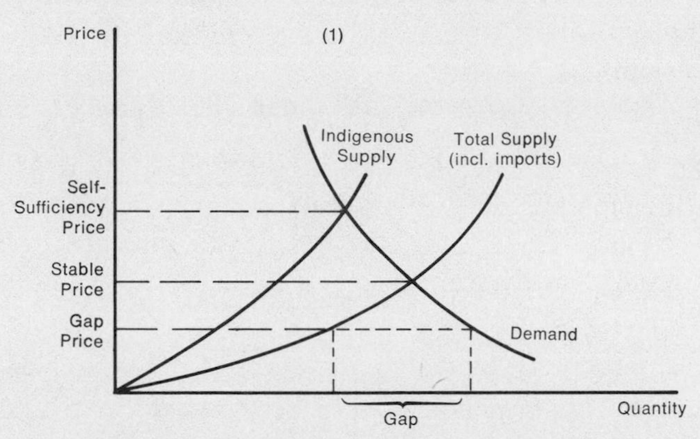

The excess of demand over indigenous supply at any price is satisfied by imports (if available).

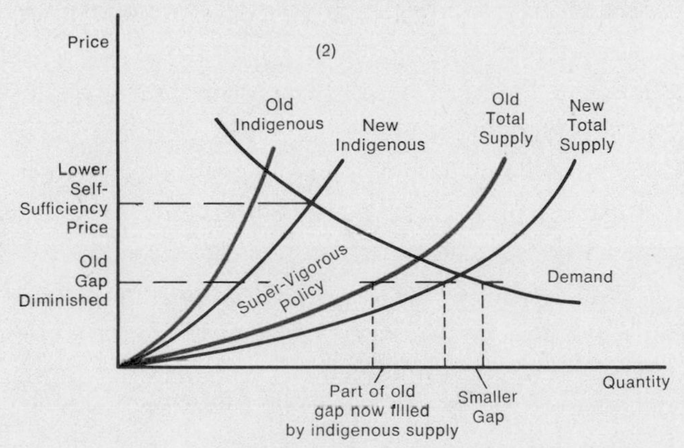

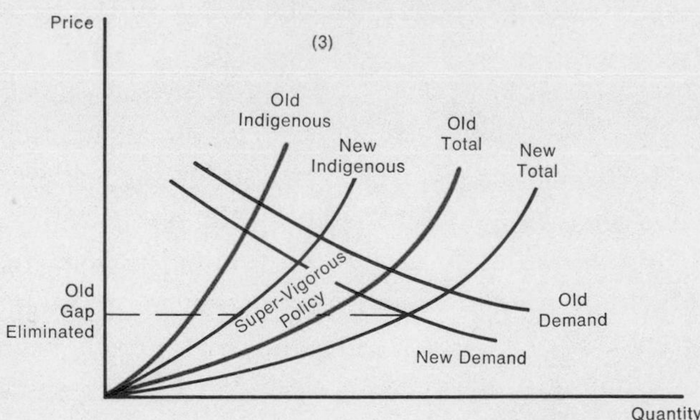

Gap closed by a combination of super-vigorous demand *and* supply policies.

actions should not be delayed if prospective shortages are to be avoided.

We recognize, of course, that by changing our assumptions—in a variety of ways—the prospective shortages in the year 2000 could be reduced or eliminated.

For example, economic growth may be forced to lower levels than we have assumed. This could have important repercussions on all countries, but the largest effects may be on the developing countries because of their dependence on trade with and aid from the industrialized regions of the world, and their inability to pay higher prices.

Another possibility for balancing energy supply and demand by 2000 is that energy prices could go higher than we have assumed. Higher prices—so long as the increases are gradual, controlled and foreseen—could encourage energy savings and stimulate alternative fuel supplies.

We have not analyzed in detail these price and growth implications. Nonetheless, detailed analyses of cases that eliminate prospective shortages would not change our consensus view—the view that the world must act on oil and gas replacement systems with a sense of urgency. Any scenario which moves the prospective shortages beyond the year 2000 only serves to delay the time when severe dislocations could occur. Such "delaying tactics" do not eliminate the problem, nor can they be implemented without great costs. To fully consider the effects beyond 2000 would require extensive analysis of supply alternatives such as solar, oil sands, and others, as well as demand factors.

If, in our Workshop, we had had time to extend our analyses beyond 2000, and more fully consider changes in assumptions, we might have been able to provide more analysis of such alternatives. We hope that this might be done by others. Yet the learning value of our present cases is high, and argues for national policies that are based on the urgent need for reducing future reliance on oil and gas.

We need not be totally restricted to the cases already described. We may ask: What if a *different* combination of WAES scenario variables were chosen for analysis? What would be the results? To answer that question, we made a special study* of Case D-3 to the

* This study was based on a highly aggregate analysis, compared to the more detailed analysis for the other 4 year 2000 WAES cases. We have not analyzed, in detail, the full implications of this case.

year 2000 (see Figure 1-4—the WAES scenario chart). Case D-3 is a low-economic-growth (2.8%/year from 1985 to 2000), rising-energy-price (to $17.25 per barrel by 2000) case. It might be labeled a "low-demand/high-supply" case—at least relative to the other WAES cases.

In Case D-3, the availability of oil and gas is probably sufficient to meet preferred demands for these fuels in the year 2000. This result assumes, of course, that our high OPEC production assumptions will occur along with the other assumptions of the case. That is merely an assumption—but a critical one; oil is still, in this case, the most important global fuel. The small margins of deficit or surplus in this case are probably well within the limits of uncertainty of the numbers themselves. Yet even if this case does balance—even if potential supplies are in fact sufficient to meet desired demands in 2000—this case just stretches the time-scale. The same mismatch between supply and demand as in other cases might appear by 2005 or 2010. This postponement might allow more time to adjust to the eventual transition, more time to develop new technologies and renewable energy forms. But these technologies must be developed and this time must be used.

Scenario Case D-3 is *assumed*, just like all others. The result is to "buy time," which is helpful only if the world uses it. That time is bought at a cost—the cost of lower economic growth (the same growth rate as the WAES "low" cases—D-7 and D-8).

The principal finding of special Case D-3 is that there is no gap. Potential oil production equals desired oil demand, at least to the year 2000. Depending on the development of other energy sources, the gap will likely first appear five to ten years after the turn of the century.*

Natural gas and coal production in the year 2000 is slightly less than the maximum potential for Case D-3. Substitutions of these fuels for oil could, conceivably, allow a slightly higher growth and lower price case to be in energy balance by 2000. For this to occur, however, natural gas and coal must be developed and traded internationally—as is the case in all WAES scenarios. And, equally important,

* A similar result—a closing of the prospective oil gap—is seen if one simply assumes high gross additions to oil reserves and a high OPEC ceiling (45 MBD) coupled with either of our low-growth, constant-price cases (D-7 or D-8). The validity and consistency of such a coupling of assumptions is open to question; nonetheless, such a combination does (just marginally) eliminate the gap.

nuclear and hydroelectric capacities must be developed to nearly the maximum levels of the case. Any feeling of relaxation derived from this case should be tempered by the recollection of those actions required to achieve the large-scale development of the several supply sources and levels of conservation already assumed in the scenario. There are no simple solutions.

Other Uncertainties

The ability to foresee that some things cannot be foreseen is essential. There are no seers who can look 10 or 25 years into the future with assured accuracy. Any study such as ours must recognize a number of categories and characteristics of uncertainties—some of which could overturn the trends in our projections. To avoid neglecting some uncertain or unknown, but possible, events of the next 25 years, we can ask what if some presently unanticipated events were to occur. We can also ask what if some events presently anticipated were *not* to occur. In either case, what might be the consequences?

Special case studies using already-defined WAES scenario variable assumptions are not the only kinds of uncertainties worth considering. The WAES methodology and analysis is based on a set of futures without large discontinuities—either major new technologies or large interruptions or disasters. Such factors cannot be included in a systematic, quantitative way in the analysis—and it is probably inappropriate to try. Nonetheless, in the real world, large unanticipated events do occur, and should not be ignored.

Some events could ease the transitions which lie ahead. Others could create very large problems.

Technological breakthroughs in a number of areas could have substantial impacts. Processes for large-scale in situ production of oil from shale, oil sands, and/or heavy oil; an increase in oil recovery significantly above 40 percent; in situ coal gasification; fusion power becoming commercially available; and rapid development and distribution of low-cost solar technologies—all could "make or break" any given set of projections. We consider it wiser to classify unanticipated bonuses of this kind as just that: bonuses. To do otherwise would be most imprudent.

On the other hand, the possibility of discontinuities and disasters cannot be ignored in good planning, either. The next 25 years could see, for example: runaway inflation or prolonged depression; local or regional wars or coups d'etat; major energy-related accidents; or terrorist activities. If any such unanticipated events were to occur, our future would be considerably different.

We do not apologize for the uncertainties in our projections. We feel our assumptions are plausible, our analyses convincing. But they do assume a "surprise-free" world. We recognize the obvious: that that may not, actually, be the case.

Conclusions

The crux of the now-familiar problem revealed by our integration analyses is that preferred demand for oil in WOCA exceeds maximum potential production of oil by an amount which grows, starting in the late 1980's at a rate of about 1.8 MBD/yr., to some 20 MBD by 2000. Using different assumptions in other cases, the size of the prospective oil gap varies from 15 to 20 MBD, and the gap shows up as early as 1981 to as late as 2004. But the same rapidly growing shortfall results. Actions are required, and on a large scale, to avoid shortages. The size of the gap is large and its inception is soon.

Among the WAES cases studied to 1985, only with low oil prices and high economic growth (Case E) is there insufficient supply to meet total energy demand within expected fuel mix preferences. In the real world, growth would be lower, energy prices would be higher than Case E, and energy policies might change—lowering demand and increasing supply, and eliminating the deficits.

While 4 of our cases "balance" in 1985, it should be remembered that the estimates assume maximum potential supplies consistent with case assumptions. Given lead times of 5-10 years or more for many projects, failure to make necessary near-term commitments or to resolve a variety of current restraints on production, or to develop future supplies may foreclose some options for 1985.

The years up to 1985 are critical ones. Events and policy decisions in the decade before 1985 will determine success in demand reduction, fuel substitution, or additions to supply in the 1985-2000 period. We are, in 1977, on the threshold of a critical decision period. We cannot afford to waste the years immediately ahead if we are to have any large-scale energy options available before the end of the century. The time for decisive action is now.

The broad directions for such action are outlined in the results of our studies of demand reduction, supply addition, and fuel substitution. These elements are built into our supply-demand integrations to the year 2000 which indicate that the energy and fuels shortfalls could be partially filled through major substitutions of fuel, if:

—nearly all fossil fuels were removed from use in electric power plants, to reduce processing losses;

—synthetic crude oil is made from available coal supplies, in order to meet essential demands for liquid fuel; and

—industrial and domestic sectors use significantly more coal and less electricity than currently and preferred patterns.

In the real world, any number of factors could combine to eliminate the gaps: growth could be lower, energy prices could be higher, or energy policies could be "super-vigorous"—lowering demand and increasing supply. But the consequences could be most undesirable. Balancing energy supply and demand by 2000 in an acceptable manner requires great effort.

The transition away from primary reliance on oil will be well under way by the year 2000. For this to be a smooth transition, greater international cooperation among increasingly interdependent nations is essential. Vigorous research, development, and demonstration of new supply sources, conservation and fuel-switching programs must move forward on an international scale. These and other implications of the year 2000 "prospective shortages" of our studies are discussed in Part I of this report.

The main thrust of our cases for the year 2000 is inescapable: the period to the end of the century will be one of *energy transition*—away from oil as the world's dominant fuel. The timing of future energy-related programs and plans must take account of the challenges of this critical period. Our energy world then and in the 21st century depends on it.

APPENDIX I

ENERGY AND ECONOMIC GROWTH PROSPECTS FOR THE DEVELOPING COUNTRIES: 1960-2000

PREFACE

The Workshop on Alternative Energy Strategies (WAES) has focused on energy supply and demand in the industrialized world. The WAES Participants come from fifteen countries, primarily from the industrialized nations that account for about 80% of the total energy consumption in the World Outside Communist Areas (WOCA). The actions of these countries—or their failure to act—to alleviate possible future shortages of world energy will significantly affect the energy prospects of all other nations.

In our analysis, it is necessary to make assumptions about the energy supply and demand patterns in what is termed "Rest of WOCA"—those countries of WOCA outside Western Europe, North America, and Japan. These countries consist primarily of the developing economies (both OPEC and non-OPEC) and also include Australia, New Zealand, and the Republic of South Africa.

We have relied extensively on others with knowledge of these countries to help analyze their future energy supply and demand prospects. In particular, individuals from the International Bank for Reconstruction and Development (World Bank) have been most helpful in estimating the developing country economic growth rates to 1990, in deriving relevant income elasticities of energy demand and in providing energy supply estimates to 1985.

The energy supply estimates by fuel type for 1975-1985 are from World Bank and WAES sources, and for 1985-2000 are taken from the WAES global supply estimates for oil, gas, and coal (see Chapters 3, 4 and 5).

The energy supply-demand estimates in this paper are very tentative. Historical data on energy consumption in developing countries is generally incomplete. A significant share of total energy comes from non-commercial sources such as firewood, cow dung, and vegetable waste, thereby confusing the data even further. This survey also attempts to cover over 90 countries and is, therefore, exceedingly general.

Our purpose is to attempt answers to the following questions which are essential to the WAES global supply-demand analysis:

1. Given certain assumptions regarding economic growth, what is the probable range of *commercial* energy consumption in the developing countries at 1985 and 2000?

2. What potential domestic energy supplies are available to help these countries meet their anticipated demands?

3. What would be the probable range of desired imports (or exports) of energy by these countries?

Attempting answers to these questions constitutes the scope of this analysis. We recognize that developing countries will need appropriate mechanisms to help them achieve desired levels of economic growth and that new arrangements may be needed to assist them in meeting the rising costs of energy. Such arrangements must be part of a much larger complex system of existing economic relationships and institutions, an analysis which is outside the scope of the WAES study.

WAES accepts full responsibility for the developing country estimates in this paper, but wishes to express its appreciation for the valuable assistance rendered by the World Bank, and in particular to Messrs. Nicholas Carter and Frank Pinto (Consultant) of the Economic Analysis and Projections Department, and to Mr. William Humphrey.

Introduction

This paper focuses on the energy and economic growth prospects of the OPEC and non-OPEC developing economies in the Rest of the

World Outside Communist Areas (WOCA).[1] The full technical report on which this summary is based is to be included in our third technical report volume—*Energy Supply and Demand Integrations to the Year 2000: Global and National Studies* (MIT Press, 1977).

Primary energy consumption in the developing countries during 1972 constituted approximately 15% of total non-Communist world energy consumption. As these countries industrialize, given certain energy-economic growth relationships, their share of total world energy consumption will rise relatively faster than that of the industrialized world. The WAES global supply-demand integrations, based in part on estimates in this paper, indicate that the developing countries could consume as much as 25% of total world energy by the year 2000, as the following pie diagram illustrates.

Projected Shares of the World's Energy Consumption

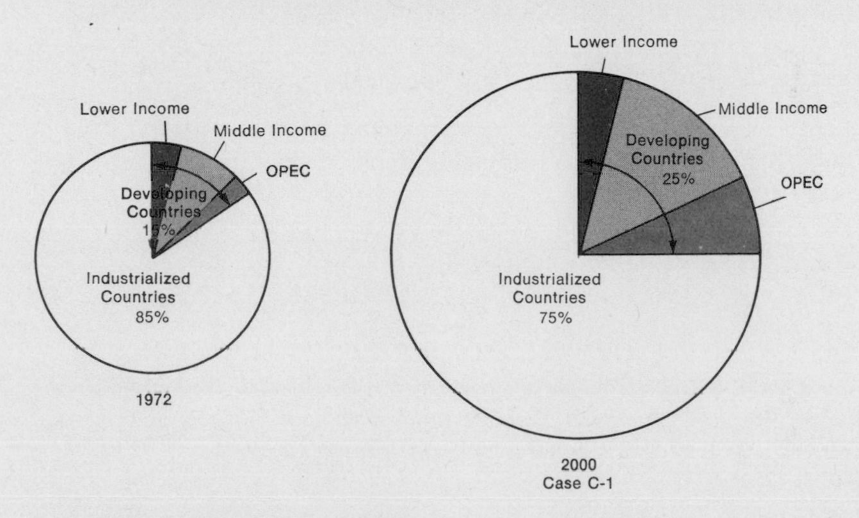

Developing Country Coverage

The ninety-three developing nations (LDCs) in WOCA considered in this analysis are divided into two major groups:

A. The 13 OPEC countries

B. The non-OPEC developing nations of Asia, Africa, and Latin America which are further divided into:

---

[1] In WAES, the "Rest of the World Outside Communist Areas" comprises the developing economies together with Australia, New Zealand, and South Africa (Rep.). A brief analysis of energy and growth prospects for the latter three nations is included in the *Energy Supply and Demand Integrations: Global and National Studies.*

i) *Lower-Income Countries*—comprising those developing economies with per capita annual income below $200. The countries in this group are located in two geographical regions, and the number of countries in each group is as follows: South Asia (7) and Lower-Income Sub-Sahara Africa (20).

ii) *Middle-Income Countries*—those with per capita annual income above $200. The countries in this group fall into three geographical regions which are (with the number of countries in each region): East Asia (9); Caribbean, Central and South America (21); Middle-Income Sub-Sahara Africa and West Asia (23). Per capita incomes range from $200 to around $1000 with the mean being about $550.[2]

These classifications are similar to those used by the World Bank. A detailed list of the countries in each major group follows:

A. *OPEC Countries*

Algeria	Iran	Libya	Saudi Arabia
Ecuador	Iraq	Nigeria	United Arab Emirates
Gabon	Kuwait	Qatar	Venezuela
Indonesia			

B. *Non-OPEC Developing Countries*

i) *Lower-Income Countries*
(annual per capita income under $200 [1972 U.S. dollars])

South Asia	*Lower-Income Sub-Sahara Africa*		
Afghanistan	Burundi	Kenya	Somalia
Bangladesh	Central African Republic	Madagascar	Sudan
Burma	Chad	Malawi	Tanzania
India	Dahomey	Mali	Togo
Nepal	Ethiopia	Niger	Uganda
Pakistan	Guinea	Rwanda	Upper Volta
Sri Lanka		Sierra Leone	Zaire

ii) *Middle-Income Countries*
(annual per capita income over $200 (1972 U.S. dollars))

East Asia	*Middle-Income Sub-Sahara Africa and West Asia*	*Caribbean, Central and South America*
Fiji	Angola	Argentina
Hong Kong	Bahrein	Barbados

[2] These distinctions are based on data in the 1974 and 1975 *World Bank Atlas*. They relate to the year 1972.

ii) *Middle-Income Countries* (Cont.)

	Middle-Income Sub-Sahara Africa and West Asia	*Caribbean, Central and South America*
Korea (South)	Cameroon	Bolivia
Malaysia	Congo P.R.	Brazil
Papua New Guinea	Cyprus	Chile
Philippines	Egypt	Colombia
Singapore	Ghana	Costa Rica
Taiwan	Israel	Dominican Republic
Thailand	Ivory Coast	El Salvador
	Jordan	Guatemala
	Lebanon	Guyana
	Liberia	Haiti
	Mauritania	Honduras
	Morocco	Jamaica
	Mozambique	Mexico
	Oman	Nicaragua
	Rhodesia	Panama
	Senegal	Paraguay
	Syria	Peru
	Tunisia	Trinidad and Tobago
	Turkey	Uruguay
	Yemen AR, DM	
	Zambia	

Population Trends

An important factor in the analysis of energy consumption in the developing world is the present level and future growth rate of population. The World Outside Communist Areas considered by WAES had a population of around 2.7 billion in 1975 and will have an estimated population of 4.5 billion by the year 2000. Of these totals, the industrialized nations of North America, Europe, Japan, and Oceania constituted approximately 28% of the total population in 1975 and will probably constitute around 20% by the year 2000.

Table 1 presents estimates of the absolute levels and growth rates of population for the developed and developing nations for the years 1970, 1975, 1985, and 2000. Most of these figures have been derived from United Nations population projections. The energy problems facing the developing economies are more severe because of their projected 2.4-2.7% average annual population growth between now and year 2000, as compared to 0.7-0.9% projected for the industrialized world.

Table 1 Estimated Developed and Developing Country Population Levels and Growth Rates: 1970-2000

	Population in Millions*				Average Annual Growth Rate of Population (percent)		
	1970	1975	1985	2000	1970-1975	1975-1985	1985-2000
Total WOCA Population	2399	2661	3310	4475	2.1	2.2	2.05
Developed Economies	702	732	792	872	0.9	0.8	0.7
Developing Economies	1697	1929	2518	3603	2.6	2.7	2.4
of which:							
(A) OPEC	255	292	388	566	2.8	2.9	2.6
(B) Non-OPEC Developing Countries	1442	1637	2130	3037	2.6	2.7	2.4
of which:							
i) Lower-Income Countries	889	1005	1301	1835	2.5	2.65	2.3
a) South Asia	740	835	1076	1487	2.4	2.6	2.2
b) Lower Income Africa	149	170	225	348	2.7	2.9	2.9
ii) Middle-Income Countries	553	632	829	1202	2.7	2.75	2.5
a) East Asia	138	158	207	290	2.8	2.7	2.3
b) Middle Income Africa and West Africa	162	184	240	353	2.6	2.7	2.6
c) Caribbean, Central and South America	253	290	382	559	2.8	2.8	2.6

* Rounded to the nearest million.

WAES Scenario Assumptions

WAES uses the scenario approach in its studies. The principal scenario variables are economic growth (high or low), real energy prices (rising, constant, or falling), government policy (vigorous or restrained), and, for 1985-2000, the principal replacement fuel (coal or nuclear). For the developing economies, only a constant real energy price to 1985 was examined. Furthermore, government policy response to 1985 is not included as a variable due to the uncertainty in modelling an aggregate policy response for over 90 countries. The WAES scenario cases considered in the analysis of developing country prospects are more fully specified in Table 2.

Methodology

The first step in our analysis is the determination of economic growth rates for the developing nations, which are derived primarily from the World Bank's SIMLINK Model.[3] High and low economic growth

[3] *The SIMLINK Model of Trade and Growth for the Developing World*, World Bank Staff Working Paper No. 220, October 1975. Also in *European Economic Review*, Vol. 7 (1976), pp. 239-255.

SIMLINK is primarily a medium-term forecasting system in which exports of the non-OPEC developing countries are related to the level of economic activity in the OECD countries through a series of individual com-

Table 2 WAES Scenario Assumptions

Case	Economic Growth	Energy Price	Principal Replacement Fuel
1976-1985			
C	High	$11.50	—
D	Low	$11.50	—
1985-2000			
C-1	High	$11.50-$17.25	Coal
C-2	High	$11.50-$17.25	Nuclear
D-7	Low	$11.50	Coal
D-8	Low	$11.50	Nuclear

rate assumptions for the developed economies, as well as the WAES oil price assumption of $11.50,[4] are special inputs to the model, whose simulations then provide economic growth projections for the major developing regions of WOCA to 1985. The following figure illustrates the SIMLINK flow from the exogenous input to the output.

The historical (1960-72) relationship between regional economic growth and energy consumption is then examined by considering the income elasticity of demand for energy use. This is defined as the growth rate of energy consumption divided by the growth rate of real income. Since real energy prices rose substantially between 1972 and 1976 and are assumed to either remain constant (WAES Cases C and D in 1985, Cases D-7 and D-8 in 2000) or increase by 50% by year 2000 (WAES Cases C-1 and C-2), the historical income elasticities were revised downwards for the period 1976-2000 to reflect the more energy-efficient use of resources in the future.

The primary energy demand growth rates are obtained from the real economic growth rates and the income elasticities of energy demand. The total primary energy consumption of developing countries in 1985 and 2000 can then be determined.

The energy supply estimates by fuel type for 1975-1985 come from

---

modity models. Growth in the developing countries is linked to investment levels and imports; imports in turn depend on export earnings and inflows of foreign capital.

Given assumptions as to OECD growth, the availability of foreign capital to the developing countries, and the international price of petroleum, the model may either be run to determine the import-constrained GDP growth rates to be anticipated in developing countries, or to determine the real resource transfer that they would need to support a specified GDP growth target.

[4] Prices in 1975 U.S. dollars per barrel of Arabian light crude oil, f.o.b. Persian Gulf.

Figure 1 Flow Diagram—SIMLINK III

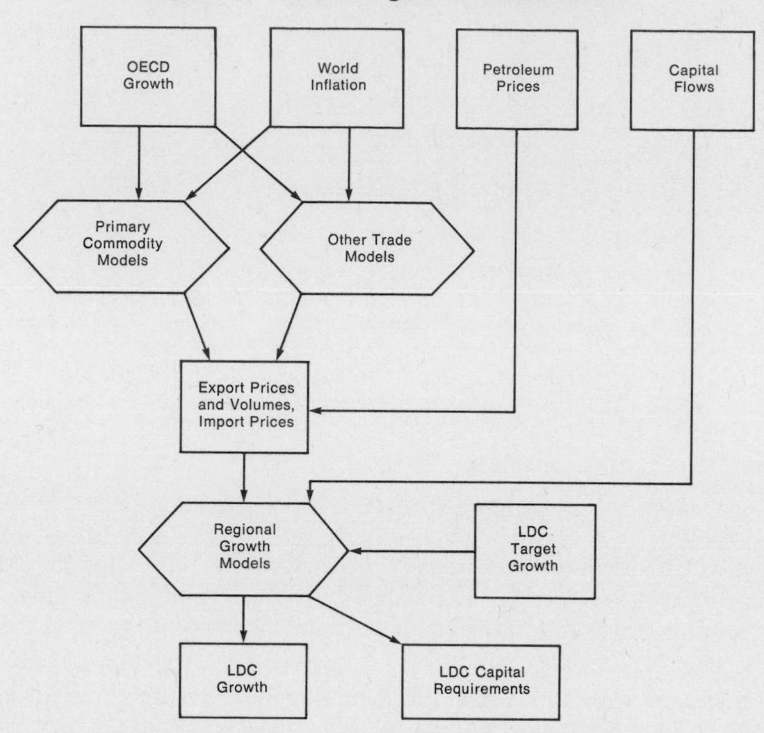

World Bank[5] and WAES sources. The 1985-2000 estimates come primarily from the WAES global supply studies described in Chapters 3 through 7.

Supply-demand integrations then balance available energy supplies with expected energy demands. The resulting figures show the range of energy exports and imports of both OPEC and non-OPEC developing countries.

Demand for Energy

Economic Growth Rate Assumptions: 1960-2000

The historical and projected economic growth rates for the developing economies for the period 1960-2000 are presented in Table 3. The 1976-1985 growth estimates have been derived primarily from simulation of the World Bank's SIMLINK Model using the WAES high (Case C) and low (Case D) economic growth assumptions for the OECD countries as well as the WAES oil price assumptions of $11.50 per barrel (1975 U.S. dollars). Because SIMLINK is primarily a medium-term forecasting sys-

[5] A. Lambertini, *Energy and Petroleum in Non-OPEC Developing Countries, 1974-80*, World Bank Staff Working Paper No. 229, February 1976.

tem, it was necessary to make extrapolations to determine the 1985-2000 economic growth rates.

Table 3 Real GNP Growth Rate Assumptions: 1972-2000

(average annual percent growth)

Period	*1960-72*	*1972-76*	*1976-1985*		*1985-2000*	
Economic Growth			*High*	*Low*	*High*	*Low*
WAES Case			*C*	*D*	*C-1,2*	*D-7,8*
Non-OPEC Developing Countries	5.6	5.1	6.1	4.1	4.6	3.6
i) Lower-Income Countries	3.7	2.3	4.4	2.8	3.1	2.5
ii) Middle-Income Countries	6.2	5.9	6.6	4.5	4.9	3.9
OPEC	7.2	12.5†	7.2	5.5	6.5	4.3
Developed Economies* (OECD)	4.9	2.0	4.9	3.1	3.7	2.5

* As derived by WAES analyses of individual countries.
† Preliminary estimate.

Several of the middle-income countries profited by the commodity boom of 1972-73 and they should be able to achieve comparatively rapid growth over the next decade. Latin American and East Asian countries are expected to show the highest economic growth patterns among the non-OPEC LDCs because of higher capital productivity, an increase in the rate of investment, and export promotion. The lower-income countries will grow much slower than the middle-income countries and they will continue to suffer from the effects of higher oil prices and agricultural shortfalls. The projected growth rates assume, however, that improvements will be achieved in domestic policies with special emphasis on increasing exports.

As a group, the non-OPEC developing countries are expected to maintain a growth rate higher than that of the OECD area. In the long run, middle-income countries are expected to grow 1-2% faster and lower-income countries about .5% slower than the developed nations. OPEC countries can be expected to achieve a high rate of economic growth during 1976-85. This high growth rate may decrease somewhat during 1985-2000 since many OPEC countries have undertaken large development projects which after 1985 may cause foreign exchange shortages and curtailed imports.

Historical Growth in Energy Consumption: 1960-72

From 1960-1972, the nations of the developing world more than doubled their consumption of commercial energy and increased their demand for electric power by more than 250%. In 1972 the developing countries accounted for around 15% of total world energy consumption in WOCA.

Sixteen countries accounted for about three-quarters of all LDC energy demand.[6] These are Argentina, Brazil, Chile, Colombia, Egypt, India, Indonesia, Iran, Korea, Mexico, Pakistan, Philippines, Taiwan, Thailand, Turkey and Venezuela.

Total commercial energy demand within the developing world (including OPEC) during 1972 was approximately 9.5 MBDOE (excluding international oil bunkers) and of this total, 25% was in lower-income countries, 58% in middle-income countries, and 17% in the OPEC countries.

Table 4 summarizes the actual quantities of commercial energy consumed by the OPEC and non-OPEC developing countries for the period 1960-72 in both metric tons of coal equivalent and in million barrels per day of oil equivalent. The energy and real income growth rates for the period are also included.

We have estimated only commercial fuel consumption. In several LDCs, noncommercial energy in the form of firewood, cow dung, and vegetable waste constitutes a significant share of total energy consumption. In India, noncommercial energy was estimated at 59% of total energy consumption in 1960, 48% in 1970, and is expected to remain a significant, though decreasing, percentage of total energy consumption.[7] Due to the great difficulty in obtaining accurate data, the noncommercial sources are excluded from this study. A detailed study on the developing world's energy prospects would have to consider such fuels explicitly.

Economic Growth and Energy Growth

Table 5 shows our assumptions regarding the historical (1960-72) and projected relationship between energy consumption and real economic growth. This relationship is termed the *income elasticity of demand for energy use.* The price of oil quadrupled between 1972 and 1976. A simple extrapolation of the historical (1960-72) income elasticity would result in an overestimation of energy demand. The much higher current and expected energy prices assumed in the WAES cases to 1985 would tend to reduce the energy intensiveness in economic activity. Specifically, the income elasticity of energy demand would decrease if the current energy prices are maintained. For the period 1985-2000, the C cases assume a 50% increase in real energy prices, and this would serve to lower the income elasticity of energy demand even further.

From the World Bank study,[8] one can derive some relationships between LDC income growth and energy consumption for the period 1970-80. In the lower-income countries, per capita income elasticity of energy

[6] *Modern Power Prospects in Developing Countries*, Richard Barber Associates (1976).
[7] *Report of the Fuel Policy Committee*, Government of India, 1974.
[8] (Refer to footnote 5).

Table 4 Developing Economy Commercial Energy Consumption and Income Elasticities of Energy Demand: 1960-72

	Energy Consumption				*Energy Growth Rate (%)*	*GNP Growth Rate (%)*	*Income Elasticity of Energy Demand*
	10^6 tons of coal equivalent		*MBDOE*				
	1960	*1972*	*1960*	*1972*	*1960-1972*	*1960-1972*	*1960-1972*
Non-OPEC Developing Countries	260.40	594.50	3.43	7.81	7.10	5.50	1.29
i) *Lower-Income Countries*	*88.50*	*177.40*	*1.17*	*2.33*	*6.00*	*3.70*	*1.62*
a) South Asia	83.40	165.10	1.10	2.17	5.80	3.60	1.61
b) Lower-Income Africa	5.10	12.30	0.07	0.16	7.60	4.00	1.90
ii) *Middle-Income Countries*	*171.90*	*417.10*	*2.26*	*5.48*	*7.70*	*6.20*	*1.24*
a) East Asia	27.80	97.60	0.37	1.28	11.00	7.70	1.43
b) Middle-Income Africa and West Asia	32.90	79.30	0.43	1.04	7.60	5.80	1.31
c) Caribbean, Central and South America	111.20	240.20	1.46	3.16	6.60	5.90	1.12
OPEC	48.30	127.90	0.64	1.68	8.40	7.20	1.17
Total Developing Economies	308.70	722.40	4.07	9.49	7.40	5.80	1.28

Sources:

Energy Consumption:

1. U.N. *World Energy Supplies, 1950-74,* Series J, No. 19.
2. India: *Report of the Fuel Policy Committee,* Government of India, 1974 and also J. D. Henderson, *India: The Energy Sector,* World Bank.
3. A. Lambertini: *Energy and Petroleum in Non-OPEC Developing Countries*, World Bank Staff Working Paper, No. 229, February 1976.
4. William Humphrey, "Estimation of Energy Growth for Non-OPEC LDCs by Region and OPEC Countries for 1972-85 and 1985-2000," Cambridge, WAES Associates Report, January 15, 1976.

GNP Data: World Bank Atlas, 1975 in constant U.S. 1973 dollars.

Conversion Factor: One MBDOE = 76 × 10^6 metric tons of coal equivalent.

Table 5 Projected LDC Income Elasticity of Energy Demand: 1960-2000

Period	1960-72	1976-85	1985-2000	
WAES Case		C & D	C-1, C-2	D-7, D-8
Oil Price	$2.00	$11.50	$11.50-$17.25	$11.50
Non-OPEC Developing Countries*	1.29	1.19	1.04	1.10
i) Lower-Income Countries	1.62	1.50	1.20	1.30
ii) Middle-Income Countries	1.24	1.10	1.00	1.05
OPEC	1.17	1.10	1.05	1.10

* The real income ratios for the Non-OPEC Developing Countries have been weighted using the following year-end weights:

Year	1972	1985		2000	
WAES Case		C	D	C-1, C-2	D-7, D-8
Lower-Income Countries	.246	.215	.219	.176	.186
Middle-Income Countries	.754	.785	.781	.824	.814
Total	1.000	1.000	1.000	1.000	1.000

demand is about 40% higher than the corresponding total income elasticity.[9] For the middle-income countries, however, per capita and total income elasticities are found to be noticeably closer. In other words, the process of industrialization that raises per capita income also results in a gradual reduction in the growth rate of energy consumption with respect to growth in real income.[10]

Table 6 specifies the percentage decrease assumed in the income elasticity of energy demand for the period 1976-1985 as compared to 1960-1972, and for 1985-2000 as compared to 1976-1985, respectively. Elasticity estimates for the lower-income countries are dominated by India, whose energy consumption in 1972 was about 75% of the total, while Argentina, Brazil and Mexico together accounted for about 42% of the energy consumption of the middle-income countries. The long-run income elasticities for the lower-income countries are consistent with those implied

[9] Per capita income elasticity $= \dfrac{\text{growth rate of energy consumption per capita}}{\text{growth rate of real income per capita}}$;

and

Total income elasticity $= \dfrac{\text{growth rate of energy consumption per capita} + \text{population growth rate}}{\text{growth rate of real income per capita} + \text{population growth rate}}$

[10] Variances in total energy elasticity with respect to GNP at different per capita income levels have been found to disappear when price-adjusted GNP is used as an independent explanatory variable in place of official exchange-rate GNPs. While attempts are being made to determine the price-adjusted GNPs for several developing nations (see the pioneering study by Kravis, Kenessey, Heston and Summers, *A System of International Comparisons of Gross Product and Purchasing Power*, 1975), this study uses readily available official exchange-rate GNPs.

in the Indian Fuel Policy Committee Study,[11] while the income elasticities for the middle-income countries are consistent with estimates provided by the WAES Mexican team.

Table 6 Percent Change Assumed in the Income Elasticity of Energy Demand over the Preceding Period*

Period	1960-72	1976-85	1985-2000	
WAES Case		C & D	C-1, C-2	D-7, D-8
Oil Price	$2.00	$11.50	$11.50-$17.25	$11.50
Non-OPEC Developing Countries		−8%	−13%	−8%
i) Lower-Income Countries		−7%	−20%	−13%
ii) Middle-Income Countries		−11%	−9%	−5%
OPEC		−6%	−5%	nil

* Percent changes rounded to nearest percent.

While the income elasticities of energy demand for non-OPEC developing countries are assumed to decline, the decrease will be more pronounced in the lower than in the middle-income countries. Their elasticities, however, will still be higher than the developed country elasticities which, in the WAES analysis, have been shown to drop below unity.

Obviously, using such a simple relationship between economic growth and energy growth is inadequate in some respects, in that it fails to include factors that may significantly affect energy consumption—such as changes in the industrial structure, or increased mechanization in agriculture. Alan Strout,[12] for instance, considers the production of a small group of key energy-intensive materials (such as iron and steel, cement, aluminum, etc.) which, when combined by using energy weights, are an indication or measure of the "energy-intensiveness" of a country's industry. For developing countries achieving rapid industrialization, the production of energy-intensive goods would have to grow even faster than normal, and income elasticities would accordingly be higher.

This raises the important issue of whether developing countries will choose to develop energy-intensive industries. Some feel a significant transfer of industry to the developing countries endowed with energy resources may occur. For example, a significant proportion of the industrialized world's aluminum processing industry might be relocated in Brazil, where abundant hydro reserves would allow for large-scale, relatively inexpensive production of aluminum. Likewise, many chemical plants might be "exported" to the OPEC countries where abundant gas reserves permit manufacturing of chemicals, such as ammonia, at lower cost. It is the assump-

[11] *Report of the Fuel Policy Committee*, Government of India, 1974.
[12] Alan Strout, *The Future of Nuclear Power in the Developing Countries*, unpublished working paper, MIT Energy Laboratory, 1976.

tion of this paper that while such transfers may occur, there will be no major transfer of energy-intensive industries from the developed to the developing world. Furthermore, the immense complications of such a transfer—both for the developed and developing countries—suggest that most developing countries will continue to be net importers of these products. Strout[13] has shown that most developing countries did not develop significant energy-intensive industries when oil prices were low, and the higher prices assumed by the WAES cases would, if anything, discourage developing countries from major investment in energy-intensive industry. This important issue needs further research and is beyond the scope of our analysis.

Estimated Developing Country Energy Consumption: 1972-2000

Table 7 shows the estimates of developing country consumption of energy for the principal WAES cases based on the analysis in the preceding sections. Non-OPEC developing countries are expected to increase their energy consumption from four to as high as five times their 1972 level by the year 2000. This is based on an average economic growth rate of between 4% and 5% per year during the period. OPEC countries are expected to achieve an even greater (500%-800%) increase in their energy consumption by the year 2000, based on an economic growth rate of between 6% and 7% per year during 1972-2000.

Table 7 Primary Energy Consumption Targets for the Developing World*

(Unit = MBDOE)

Year	*1972*	*1985*		*2000*	
WAES Case		*C*	*D*	*C-1, C-2*	*D-7, D-8*
Non-OPEC Developing Countries	7.81	18.20	14.90	35.60	26.50
i) Lower-Income Countries	2.33	4.90	4.00	8.40	6.50
ii) Middle-Income Countries	5.48	13.30	10.90	27.20	20.00
OPEC	1.68	4.90	4.20	13.10	8.40
All Developing Nations	9.49	23.10	19.10	48.70	34.90

* Excludes bunkers.

Supply of Energy

Estimated Developing Country Energy Supplies: 1972-2000

The OPEC and non-OPEC developing countries have the potential to increase significantly their domestic energy supplies by the year 2000. There are a number of reasons. First, the higher energy prices will en-

[13] (See footnote 12).

courage the exploration for and production of fuels in many developing countries where such activities were seen as uneconomic before. This is particularly true for oil, natural gas, and coal. Developing countries with adequate reserves of coal and natural gas may choose to develop these rather than risk further deterioration in their balance of payments because of higher oil import bills.

It is difficult to generalize for over 90 countries. Therefore, in discussing energy supply we shall distinguish between OPEC and non-OPEC developing economies. Even within these two groups differences are substantial. OPEC countries differ widely in their economic growth potentials, development plans, population, and need for revenue. The non-OPEC LDCs also vary greatly in their energy supply potentials and revenue needs. Some, such as Mexico and Brazil, will most likely attain a high degree of energy self-sufficiency, while others, particularly the lower-income countries of Africa and Asia, will continue to depend on energy imports.

Table 8 shows the resource base for the non-OPEC developing countries by middle- and lower-income groupings. The middle-income countries are relatively well endowed with energy resources—especially oil and natural gas. The lower-income countries do not have abundant supplies of oil and natural gas but they do have plentiful coal reserves. These resource statistics indicate that, as a group, the developing countries have sufficient reserves of the major fossil fuels. Table 9 is a summary table of the energy production potential of the developing countries by fuel type.

Table 8 Non-OPEC Developing Countries' Medium-Term Energy Resources

(Million metric tons of oil equivalent)

	Oil[1]	*Natural Gas*[1]	*Coal*[2]	*Hydro Power*[3]	*Nuclear Power*[2]	*Total*
Middle Income	3,480	1,835	5,320	190	350	11,175
Lower Income	180	480	9,250	185	1,050	11,145
Total:	3,660	2,315	14,570	375	1,400	22,320

1 Economically recoverable at current prices and costs.
2 The measured or reasonably assured fraction of resources which could be economically exploited in the coming 5 years.
3 All estimated reserves.

Source: Energy and Petroleum in Non-OPEC Developing Countries: 1974-80, World Bank Staff Working Paper No. 229, February 1976.

Oil

(a) *OPEC Oil Production*

OPEC countries currently account for about 80% of WOCA oil

Table 9 Projected Production of Primary Energy for All Developing Economies: 1985-2000
Unit: MBDOE

	1985		2000			
	Case C	*Case D*	*Case C-1*	*Case C-2*	*Case D-7*	*Case D-8*
Scenario Assumptions						
Economic Growth	High	Low	High	High	Low	Low
Oil/Energy Price*	$11.50	$11.50	$11.50-17.25	$11.50-17.25	$11.50	$11.50
Principal Replacement Fuel			Coal	Nuclear	Coal	Nuclear
Oil						
Non-OPEC Developing Countries	6.70	5.10	11.50	11.50	7.00	7.00
OPEC	40.00	36.00	45.00	45.00	39.00	39.00
Total	46.70	41.10	56.50	56.50	46.00	46.00
Solid Fuels						
Non-OPEC Developing Countries	2.67	2.24	5.26	5.00	4.10	3.60
OPEC	.42	.30	1.46	1.30	.80	.80
Total	3.09	2.54	6.72	6.30	4.90	4.40
Natural Gas						
Non-OPEC Developing Countries	2.40	2.10	4.56	4.50	3.88	3.40
OPEC	5.47	5.08	13.60	11.60	12.30	10.20
Total	7.87	7.18	18.16	16.10	16.18	13.60
Hydroelectricity						
Non-OPEC Developing Countries	1.70	1.80	3.80	3.10	2.70	2.10
OPEC	.10	.10	.40	.40	.30	.30
Total	1.80	1.90	4.20	3.50	3.00	2.40
Nuclear Electricity						
Non-OPEC Developing Countries	1.10	.40	4.00	8.30	2.80	5.90
OPEC	.10	.10	.46	1.00	.30	.60
Total	1.20	.50	4.46	9.30	3.10	6.50
Total Primary Energy						
Non-OPEC Developing Countries	14.57	11.64	29.12	32.40	20.48	22.00
OPEC	46.09	41.58	60.92	59.30	52.70	50.90
Total	60.66	53.22	90.04	91.70	73.18	72.90

* 1975 U.S. dollars per barrel of Arabian light crude oil, f.o.b. Persian Gulf.

reserves—about 450 billion barrels. OPEC production in 1975 was about 27 million barrels a day (MBD), most of which was exported to Western Europe, Japan, and North America. Because OPEC oil production is one of the most critical elements of world energy prospects, WAES has prepared a major study on OPEC oil production potential, described in Chapter 3 of this report.

(b) *Non-OPEC Oil Production*

Proven oil reserves in non-OPEC developing countries are equal to about 26 billion barrels—5% of total WOCA reserves. However the region is vast and is to a large extent relatively unexplored for oil. Thus, extensive exploration and development, encouraged by a desire to reduce dependence on imported oil, could lead to a significant increase in production.

We therefore make two assumptions regarding discovery rates for the region. A high future discovery rate is assumed to be 4 billion barrels a year. This would permit production to more than triple today's levels —reaching 11.5 MBD by year 2000. A less optimistic discovery rate (around 2 billion barrels a year) would probably allow oil production to be twice as high as current levels by year 2000.

Much of this production is likely to be limited to a few countries. Walter Levy notes that three countries—Mexico, Brazil, and Egypt—could account for as much as 40% of total non-OPEC developing country oil production by 1985.[14] Production in Mexico alone could be 25% of the non-OPEC developing country production by year 2000. Thus, although production for the non-OPEC LDCs, on total, is likely to increase, there is still the problem of the non-oil producing LDCs who will have to continue to face large oil import bills and corresponding balance of payments deficits.

Natural Gas

(a) *OPEC Gas Production*

Proven reserves of natural gas in OPEC are estimated to be 140 billion barrels of oil equivalent—about 60% of WOCA reserves. There is, at present, only a limited demand within OPEC for natural gas, though future consumption is expected to increase substantially. Future natural gas production will primarily depend on two factors:

—Domestic requirements for natural gas (including gas for reinjection into oil wells to improve recovery rates); and

—Expected level of exports to major consuming countries.

---

[14] *OPEC in the Medium-Term*, W.J. Levy Consultants, New York, September 1976.

The levels of possible OPEC gas production have been discussed further in Chapter 4 on natural gas.

(b) *Non-OPEC Gas Production*

Despite natural gas reserves of approximately 17 billion barrels oil equivalent, current natural gas consumption in non-OPEC LDCs is small. The World Bank reports that only about 65% of the produced gas in 1973 was marketed, with the remainder either flared, vented, or reinjected. Reserves have largely remained undeveloped due to the lack of markets. With rare exception, natural gas has been produced only to meet export demand and for use in maintaining pressure in oil fields. Afghanistan, Bolivia and Brunei have been small exporters of natural gas.

There are indications that gas consumption in this region is rising. Between 1960 and 1974, its share in total energy consumption rose from 4% to 8%. It seems plausible from a resource perspective that natural gas production and consumption will continue to increase. By the year 2000, the resource base would enable production in non-OPEC LDCs to reach 4.5 MBDOE. This level would satisfy domestic consumption needs as well as allow for some export.

Solid Fuels

Coal deposits have been found in many countries, but the largest known reserves are concentrated in comparatively few areas, notably the U.S.A., the U.S.S.R., and China. Most of the coal discovered so far lies in the northern temperate zone due to extensive exploration. Few of the developing countries have ever had to look for coal, since historically their development processes got under way after oil was generally available. However, there could be significant coal reserves in many parts of the developing world. Rising oil prices may encourage non-OPEC developing countries to increasingly look toward coal to help meet domestic energy needs, and, in certain cases, possibly contribute to exports. Chapter 5, Coal, discusses the potential for developing countries' coal production.

Electrical Generation

Electrical capacity nearly tripled in the developing world during the period 1960-1973, and this rapid growth is expected to continue into the future. Over the next five years, the World Bank estimates that electrical generation will increase another 50% in the non-OPEC developing countries.[15]

Table 10 summarizes electricity growth by fuel type for the WAES scenario cases. Primary electricity use is expected to increase from 2.2 MBDOE in 1972 to 5.8 MBDOE in 1985 in the high growth case (Case C).

[15] World Bank Staff Working Paper No. 229.

Table 10 Primary Electricity Generation in Developing Countries: 1972-2000

Year	*1972*	*1985*		*2000*			
WAES Case		*Case C*	*Case D*	*Case C-1*	*Case C-2*	*Case D-7*	*Case D-8*
Scenario Assumptions							
Economic Growth		High	Low	High	High	Low	Low
Energy Price		$11.50	$11.50	$11.50-17.25	$11.50-17.25	$11.50	$11.50
Principal Replacement Fuel				Coal	Nuclear	Coal	Nuclear
Primary Electrical Generation							
Of Which:	2.21	5.82	4.81	14.28	16.53	10.20	11.57
Oil	.81	1.13	.95	2.85	1.66	2.00	1.16
Gas	.13	.63	.53	1.28	.47	.57	.35
Coal	.29	1.13	.95	1.45	1.66	1.53	1.16
Hydro	.97	1.76	1.85	4.24	3.50	3.00	2.40
Nuclear	.01	1.17	.53	4.46	9.24	3.10	6.50
Percent Primary Electrical Generation							
Of Which:	100	100	100	100	100	100	100
Oil	37	20	20	20	10	20	10
Gas	6	11	11	9	3	6	3
Coal	13	20	20	10	10	15	10
Hydro	44	31	38	30	21	29	21
Nuclear	—	18	11	31	56	30	56

By the year 2000, total installed electrical capacity in a high nuclear future could be as high as 16.5 MBDOE (730 installed GW(e)).[16] The International Atomic Energy Agency estimates that about 11% of the total generating capacity in developing countries would be in OPEC countries and 89% in non-OPEC countries.[17]

Most of this increased demand for electricity will be met by expanded hydroelectric and nuclear capacity. Reserves of hydroelectric power are abundant in developing countries, and only 4% of current potential is being utilized.[18] In 1972, hydroelectricity constituted about 44% of the non-OPEC developing countries' electrical generation and there are plans to greatly expand it. Hydroelectric capacity in developing countries is projected to increase over fourfold by year 2000 in one of the WAES cases.

Similarly, nuclear power is expected to grow rapidly. The maximum potential nuclear installed capacity for the developing countries (OPEC and non-OPEC) in year 2000 could be as high as 416 GW(e). This high estimate is from the OECD/IAEA survey of December 1975, but there are now indications that installed capacity may be considerably less. All other cases (having scenario assumptions of low nuclear growth, low economic growth or both) are scaled downward. Nuclear issues were further explored in Chapter 6.

Energy Supply-Demand Integration

We have discussed in previous sections the results of our analysis on future energy demand and supply in the developing countries. The interactions of energy demand and energy supply are complex. WAES has developed methods of comparing desired demands and potential supplies. These are our supply-demand integrations. The techniques used in the developing country analysis are simpler than those used in other national and global integrations. We have assumed certain shares of energy demand for various sectors—transport, industry, domestic/commercial, and non-energy uses. And we have further made the simplifying assumption that energy use in all sectors will grow at about the same rate as total energy demand. These assumptions are shown in Table 11. Fuel mix assumptions to year 2000 for the various sectors are shown in Table 12. Oil is expected to constitute almost all transportation's requirements in the cases. The industrial sector's fuel needs are met by substantial amounts of oil and gas, some coal, and, in the later years, some electricity. The relatively large

---

[16] 100 GW(e) of installed capacity = 2.4 MBDOE primary energy assuming 35% efficiency and a 60% load factor.

[17] International Atomic Energy Agency, *Market Survey for Nuclear Power in Developing Countries*, 1974.

[18] World Bank Staff Working Paper No. 229.

share of gas in the industrial sector is due largely to the possibility of extensive use of gas by OPEC countries. The domestic and commercial sectors use substantial amounts of oil and electricity. Non-energy demands for feedstocks and asphalt are made up largely of oil and some gas. Table 13 shows the total amount of energy consumed in each of the WAES cases when these fuel mix assumptions are incorporated.

Table 11 Developing Countries' Sector Share of Final Energy Demand: 1972-2000*

	1972	*1985*	*2000*
Transport	34%	35%	34%
Industrial	42%	42%	41%
Domestic/Commercial	18%	16%	17%
Non-Energy	6%	7%	8%

* Excluding processing losses

Table 12 Developing Countries' Fuel Mix Assumptions by Demand Sector (percent)

	Oil	*Coal*	*Gas*	*Electricity*
1972				
Transport	91	9	—	—
Industry	48	26	15	11
Domestic/Commercial	66	9	8	17
1985: C, D Cases				
Transport	97	2	—	1
Industry	42	20	24	14
Domestic/Commercial	61	3	13	23
2000: C-1 Case				
Transport	95	2	—	3
Industry	35	17	31	17
Domestic/Commercial	55	5	15	25
2000: C-2 Case				
Transport	95	2	—	3
Industry	37	16	28	19
Domestic/Commercial	55	3	15	27
2000: D-7 Case				
Transport	95	2	—	3
Industry	31	17	36	16
Domestic/Commercial	55	5	15	25
2000: D-8 Case				
Transport	95	2	—	3
Industry	37	18	27	18
Domestic/Commercial	55	3	15	27

Imports and Exports

By comparing the tables on production (Table 9) and consumption (Table 13), desired imports and potential exports by fuel type can be determined for OPEC and non-OPEC LDCs. Table 14 shows these balancing calculations.

(a) *OPEC Exports*

OPEC countries will continue to be large exporters, particularly of oil and gas. The potential varies among OPEC countries. Saudi Arabia and other Arabian Peninsula countries have the potential to maintain high export levels. Other OPEC countries, such as Venezuela and Indonesia, have growing domestic needs for their resources, which limit their exports. Total OPEC oil exports could range between 33 and 37 MBD in 1985 and between 35 and 38 MBD in 2000. The year 2000 estimates are based on OPEC production ceilings described in Chapters 3 and 8.

OPEC countries with plentiful gas reserves could export large quantities of gas but this depends on whether they are willing to increase exports significantly and on whether the importing countries will construct the expensive and complicated liquified natural gas (LNG) systems needed. By 1985, maximum planned and potential natural gas imports from OPEC are estimated to be about 3.5 MBDOE. If all planned projects mature, about 2.5 MBDOE could be delivered as LNG. This would require OPEC production of 3.3 MBDOE for export allowing for the 25% losses in the processing of LNG. Another .6 MBDOE could be delivered by pipeline from North Africa and the Middle East, resulting in a total OPEC capacity of 3.9 MBDOE of gas exports. As stated in the main report, if all import demands of the major consuming countries are to be met in year 2000, OPEC countries would need to produce as much as 10 MBDOE for export in addition to their internal demand requirements of 5 MBDOE in Case C-1.

(b) *Non-OPEC LDC Imports and Exports*

Non-OPEC LDCs have traditionally been large importers of energy though their dependence on imports varies. During the period 1960-1972 they imported as a group about 30% of their energy needs. Our projections show that in 1985 they will have to import 18-22% and by year 2000, 15-25% of their energy requirements, depending on the scenarios assumed.

The non-OPEC LDCs differ widely in their dependence on imports. Some developing countries such as Mexico and Brazil may be able to reach energy self-sufficiency. But, most of the non-OPEC LDCs will continue to depend on imports—especially of oil. These countries may require oil imports of as much as 4 MBD in 1985 and between 7.6 MBD and 9.4 MBD in the year 2000.

Table 13 Projected Developing Countries' Consumption of Primary Energy: 1985-2000
Unit: MBDOE

	1985		2000			
	Case C	*Case D*	*Case C-1*	*Case C-2*	*Case D-7*	*Case D-8*
Scenario Assumptions						
Economic Growth	High	Low	High	High	Low	Low
Oil/Energy Price*	$11.50	$11.50	$11.50-17.25	$11.50-17.25	$11.50	$11.50
Principal Replacement Fuel			Coal	Nuclear	Coal	Nuclear
Oil						
Non-OPEC Developing Countries	10.70	8.94	20.87	19.15	15.20	14.60
OPEC	3.30	2.76	6.33	7.80	3.85	4.50
Total	14.00	11.70	27.20	26.95	19.05	19.10
Solid Fuels						
Non-OPEC Developing Countries	2.40	2.11	4.03	4.19	3.63	3.20
OPEC	.42	.30	.46	.46	.20	.20
Total	2.82	2.41	4.49	4.65	3.83	3.40
Natural Gas						
Non-OPEC Developing Countries	1.92	1.63	3.56	3.50	2.88	2.40
OPEC	1.54	1.15	5.00	3.97	3.42	2.76
Total	3.46	2.78	8.56	7.47	6.30	5.16
Hydroelectricity						
Non-OPEC Developing Countries	1.70	1.80	3.80	3.10	2.70	2.10
OPEC	.10	.10	.40	.40	.30	.30
Total	1.80	1.90	4.20	3.50	3.00	2.40
Nuclear Electricity						
Non-OPEC Developing Countries	1.10	.40	4.00	8.30	2.80	5.90
OPEC	.10	.10	.46	1.00	.30	.60
Total	1.20	.50	4.46	9.30	3.10	6.50
Total Primary Energy**						
Non-OPEC Developing Countries	17.82	14.88	36.26	38.24	27.21	28.20
OPEC	5.46	4.41	12.65	13.63	8.07	8.36
Total	23.28	19.29	48.91	51.87	35.28	36.56

* 1975 U.S. dollars per barrel of Arabian light crude oil, f.o.b. Persian Gulf.
** Totals are from supply-demand integration worksheets and may not exactly equal the targets in Table 7.

Table 14 Projected (Imports) and Exports of Primary Energy for All Developing Economies: 1985-2000
Unit: MBDOE

	1985		*2000*			
	Case C	*Case D*	*Case C-1*	*Case C-2*	*Case D-7*	*Case D-8*
Scenario Assumptions						
Economic Growth	High	Low	High	High	Low	Low
Oil/Energy Price*	$11.50	$11.50	$11.50-17.25	$11.50-17.25	$11.50	$11.50
Principal Replacement Fuel			Coal	Nuclear	Coal	Nuclear
Oil						
Non-OPEC Developing Countries	(4.00)	(3.84)	(9.37)	(7.65)	(8.20)	(7.60)
OPEC	36.70	33.24	38.67	37.20	35.15	34.50
Total	32.70	29.40	29.30	29.55	26.95	26.90
Solid Fuels						
Non-OPEC Developing Countries	.27	.13	1.23	.81	.47	.40
OPEC	0	0	1.00	.84	.60	.60
Total	.27	.13	2.23	1.65	1.07	1.00
Natural Gas						
Non-OPEC Developing Countries	.48	.47	1.00	1.00	1.00	1.00
OPEC	3.93	3.93	8.60	7.63	8.88	7.44
Total	4.41	4.40	9.60	8.63	9.88	8.44
Hydroelectricity						
Non-OPEC Developing Countries						
OPEC						
Total						
Nuclear Electricity						
Non-OPEC Developing Countries						
OPEC						
Total						
Total Primary Energy						
Non-OPEC Developing Countries	(3.25)	(3.24)	(7.14)	(5.84)	(6.73)	(6.20)
OPEC	40.63	37.17	48.27	45.67	44.63	42.54
Total	37.38	33.93	41.13	39.83	37.90	36.34

* 1975 U.S. dollars per barrel of Arabian light crude oil, f.o.b. Persian Gulf.

As described earlier, certain countries could produce substantial coal and gas. Therefore the non-OPEC LDCs as a group will require no imports of coal or gas, and in fact may be modest exporters. However, this statistic masks the fact that, as in the case of oil, production is likely to come from only a few countries.